# Ergonomics in Computerized Offices

# Ergonomics in Computerized Offices

## Etienne Grandjean

*Former Director of the Department of Ergonomics and Hygiene, Swiss Federal Institute of Technology, Zürich*

*Taylor & Francis*
*London • New York • Philadelphia*
*1987*

| UK | Taylor & Francis Ltd, 4 John St, London WC1N 2ET |
| USA | Taylor & Francis, 1900 Frost Road, Suite 101, Bristol, PA 19007 |

Copyright © E. Grandjean 1987

Reprinted 1990, 1992

*All rights reserved. No part of this publication may be reproduced, stored in a retrieval system, or transmitted, in any form or by any means, electronic, electrostatic, magnetic tape, mechanical, photocopying, recording or otherwise, without the prior permission of the copyright owner and publishers.*

**British Library Cataloguing in Publication Data**

Grandjean, E.
   Ergonomics in computerized offices.
   1. Video display terminals—Human factors
   I. Title
   004.7'7        TK7882.16

ISBN 0-85066-349-0
ISBN 0-85066-350-4 Pbk

**Library of Congress Cataloging in Publication Data**

Grandjean, E. (Etienne)
   Ergonomics in computerized offices.

   Bibliography: p.
   Includes index.
   1. Office practice—Automation—Psychological aspects. 2. Video display terminals—Hygienic aspects. 3. Human engineering.    I. Title.
HF5547.5.G693  1987    651.8'01'9    86-14588
ISBN 0-85066-349-0
ISBN 0-85066-350-4 (pbk.)

∞ The paper in this publication meets the requirements of the ANSI Standard Z39.48-1984 (Permanence of Paper)

*Cover design by Russell Beach*
*Typeset by Mathematical Composition Setters Ltd, Salisbury*
*Printed in Great Britain by Taylor & Francis (Printers) Ltd, Basingstoke, Hants.*
*Second reprinting in the United States by Braun-Brumfield, Inc.*

# *Foreword*

## Harry L. Davis

*President of the International Ergonomics Association, past President of the Human Factors Society and co-founder of one of the earliest Ergonomics Practitioner Organizations in industry, at the Eastman Kodak Company.*

Other than the invention of the steam engine by Watt and the consequent mechanization and industrialization of human work, perhaps no single technological advancement in how work is organized and performed has caused as much concern among humans, and their social and technical organizations, as the invention and subsequent proliferation of the computer. The computer is totally changing the way humans collect and manipulate data. It has ushered in the information age. In offices, it is completely changing the way work and work groups are organized. It is changing the very workplace in which people perform their daily work; and it is that workplace, that working environment, and those working humans that this book addresses.

With the extensively researched and factual information that is presented in this book, ergonomists and engineers can design the working environment and the workplace to conform to human needs and human capabilities. Thus will be created a work situation that is more comfortable, pleasant, productively efficient and less stressful.

The author of this book, Professor Etienne Grandjean, was one of the early pioneers in researching human capabilities and limitations within industry, and applying that knowledge to the design of industrial work situations. He has also been a pioneer in researching and applying these human requirements to the design of automated offices. This book is an important contribution for the practitioners of ergonomics worldwide, as have many books Professor Grandjean has previously published. Workers worldwide will benefit from the improved design of automated offices brought about by the use of data and methods contained herein.

# Contents

| | | |
|---|---|---|
| **Foreword** *by Professor Harry Davis* | | v |
| **1** | **The present metamorphosis of offices** | 1 |
| **2** | **VDT jobs seen through ergonomic-tinted spectacles** | 6 |
| **3** | **Physical characteristics of VDTs** | 10 |
| **4** | **Vision** | 16 |
| 4.1. | The visual system | 16 |
| 4.2. | Accommodation | 19 |
| 4.3. | The aperture of the pupil | 22 |
| 4.4. | The adaptation of the retina | 23 |
| 4.5. | Eye movements | 24 |
| 4.6. | Visual capacities | 25 |
| 4.7. | Physiology of reading | 28 |
| **5** | **Ergonomic principles of lighting in offices** | 32 |
| 5.1. | Light measurement and light sources | 32 |
| 5.2. | Illumination level | 36 |
| 5.3. | Spatial balance of surface luminances | 40 |
| 5.4. | Temporal uniformity of lighting | 46 |
| 5.5. | Appropriate lights | 47 |
| **6** | **Visual strain and photometric characteristics of VDTs** | 55 |
| 6.1. | Eye complaints of VDT operators | 55 |
| 6.2. | Photometric characteristics of displays | 65 |
| 6.3. | Equipment and methods to measure photometric qualities of VDTs | 66 |
| 6.4. | Oscillating luminances of characters | 69 |
| 6.5. | Sharpness of characters | 73 |
| 6.6. | Character contrasts | 76 |
| 6.7. | Stability of characters | 79 |
| 6.8. | Reflections on screen surfaces | 82 |
| 6.9. | Size of characters and typeface | 86 |
| 6.10. | Dark versus bright characters | 91 |

| | | |
|---|---|---|
| 7 | **Ergonomic design of VDT workstations** | 96 |
| 7.1. | Constrained postures are long-lasting static efforts for the muscles involved | 96 |
| 7.2. | Body size and the design of workstations for traditional office jobs | 101 |
| 7.3. | Field studies on musculoskeletal troubles of office employees | 107 |
| 7.4. | Postures, workstation characteristics and physical discomfort | 112 |
| 7.5. | Orthopaedic aspects of the sitting posture | 119 |
| 7.6. | Ergonomic design of office chairs | 130 |
| 7.7. | VDT workstation design: preferred settings and their effects | 135 |
| 7.8. | The VDT keyboard | 150 |
| 7.9. | Arrangement of work surfaces in computerized offices | 154 |
| | | |
| 8 | **Noise** | 157 |
| | | |
| 9 | **Occupational stress, work satisfaction and job design** | 167 |
| 9.1. | Occupational stress | 167 |
| 9.2. | Job satisfaction versus boredom | 170 |
| 9.3. | Alleged stress among VDT operators | 173 |
| 9.4. | Job design | 185 |
| | | |
| 10 | **Radiation, electrostatic fields and alleged health hazards** | 193 |
| 10.1. | Electromagnetic radiation emission from VDTs | 193 |
| 10.2. | Electrostatic fields | 196 |
| 10.3. | Skin rashes | 198 |
| 10.4. | Alleged cataracts | 200 |
| 10.5. | Alleged reproductive hazards due to VDT work | 202 |
| | | |
| 11 | **Recommendations for VDT workstations** | 206 |
| 11.1. | Lighting | 206 |
| 11.2. | Photometric qualities of VDTs | 207 |
| 11.3. | Ergonomic design of office furniture and keyboards | 208 |
| 11.4. | Job design for VDT operators | 209 |
| | | |
| **References** | | 211 |
| | | |
| **Index** | | 223 |

**Author's acknowledgement**

My thanks are due to Frau Ilse Neur-Fannenböck for correcting my english and typing the manuscript.

# 1. The Present Metamorphosis of Offices

*A retrospective glance*

The tremendous growth in computing and the whole area of information technology has already been well described in several books and many journals. In fact, developments and implementation in this area are very recent. Stewart (195) reports that one of the earliest computers used for commercial administration in England was the 'Lyons Electronic Office', which started regular work for the Lyons Tea Shop Company way back in 1952. In the first period, lasting about 15–20 years, all information was entered into the computer through punch cards and the results were printed on paper. This rather slow procedure was replaced by keyboards and cathode ray tubes (CRTs). The keyboards greatly increased the speed of feeding information into the computer and the cathode ray tube — the main element of the visual display terminal (VDT) — gave the results immediately to the operator.

In the same period the first three generations of basic components of the computer were developed: the valve, the transistor and the large-scale integrated chip. A fourth generation, very large-scale integration, is already well advanced. This enormous growth in power and speed as well as the reduction in size and cost have led to a tremendous spread of computer machines and information systems. And so the 'age of information' is beginning.

Parallel to the development of basic components a rapid change in office technology is evident. New generations of equipment seem to spring up nearly every month, each offering leaps forward in performance and ease of use.

*A glimpse of the future*

The US National Academy of Science (153) estimated that the number of VDT operators in the United States was approximately 7 million in 1980, with about 5–10 million terminals in use.

According to Korell (108), the growth of VDTs in offices will continue. He anticipates that VDT use itself will begin to taper off in the next decade as portable and flat screen devices equipped with liquid crystal displays come into their own, but

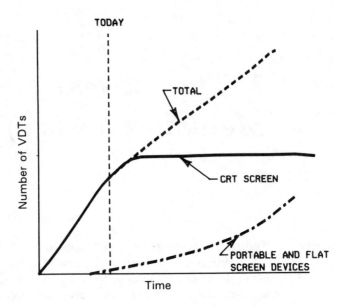

**Figure 1 Trends in growth of VDTs in offices.**
*VDT use will gradually slow down as new display types become available, but total display use will continue to grow. According to Korell (108).*

the overall penetration of computer displays into the office will continue to grow sharply for the foreseeable future. Figure 1 shows these trends.

The traditional keyboard will to some extent be supplemented by touch pads, activated surfaces and voice entry. Office noise might become a new delicate problem, since voice entry will require lower noise levels.

*The information age*

The 'industrial age', beginning in the 19th century, increased the physical power of mankind as the machines enhanced or replaced human muscular capabilities. The physical strain of human work was greatly reduced. The information age is having similar effects on mental work: information technology can enhance or replace mankind's intellectual powers and capabilities. Communications hardware and techniques are becoming vastly more complex and a critical element of the office. Fibre optics will increasingly connect computer networks and other communications hardware. These technologies are likely to continue to flood offices at an accelerating rate. There will be less centrally located, shared equipment and more individual workstations, which can already be observed today with the widely used personal computers. Most probably the number of 'knowledge' workers will increase and fewer clerical and support workers will be required.

It is unlikely that ergonomics will become redundant in the

office of the information age. In general, experience has shown that with increasing productivity the intensity of human work increases. The load on the sensory organs and mental functions, environmental problems and constrained postures are likely to remain challenges for ergonomics in the future, too.

*Traditional working conditions*

As already mentioned, VDTs are at present invading all types of offices. They are entering a world where machines have not been used before. The result is a considerable change to offices and in working conditions. To call the present change a metamorphosis, similar to that known for caterpillars and butterflies, is hardly an exaggeration.

At the traditional office desk an employee performs a great variety of physical and mental activities and has a large space for various body postures and movements: he/she might look for documents, take notes, file correspondence, use the telephone, read, exchange information with colleagues, or type for a while, and he/she will leave the desk many times during the course of the working day. Figure 2 illustrates the variety of activities in a traditional office job. A desk which is too low or too high, an unfavourable chair, insufficient lighting conditions or other ergonomic shortcomings are not likely to cause annoyance or physical discomfort. The wide choice of activities greatly reduces or precludes the adverse effects of continuous physical or mental loads of long duration.

*Man—machine systems*

The situation is, however, entirely different for an operator working with a VDT for several hours without interruption or perhaps for a whole day. *Such a VDT operator is tied to a man—machine system.* His/her movements are restricted, attention is concentrated on the screen and the hands are linked to the keyboard. Figure 3 illustrates this link between the operator and the VDT workstation. These operators are more vulnerable to ergonomic short-comings, they are more susceptible to constrained postures, poor photometric display characteristics and inadequate lighting conditions. This is the reason why the computerized office needs ergonomics — the VDT workstation has become the launch vehicle for ergonomics in the office world.

*Reports on discomfort*

For as long as engineers and other highly motivated experts operated VDTs, nobody complained about negative effects. However, the situation changed drastically with the expansion of VDTs to workplaces where traditional working methods had formerly been applied: complaints from VDT operators about visual strain and physical discomfort in the neck—shoulder area and in the back became more and more frequent. This has provoked differing reactions: some believe that the complaints are highly exaggerated and mainly a pretext for social and political claims, while others consider

4  *Ergonomics in Computerized Offices*

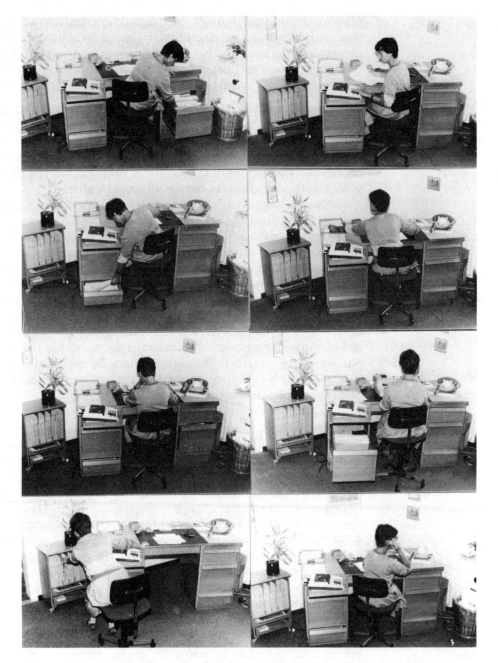

Figure 2 At traditional office desks employees carry out many different tasks and therefore have a large space for various body postures and movements.

the complaints to be symptoms of a health hazard requiring immediate measures to protect operators from injuries to their health. *Ergonomics as a science stands between these opposite poles; its duty is to analyse the situation objectively and to deduce guidelines for the appropriate design of VDT workstations. This is also the main purpose of this book.*

**Figure 3 The VDT operator is tied to a man–machine system.**
*Attention is concentrated on the screen, the hands are linked to the keyboard; constrained postures are inevitable.*

# 2. VDT Jobs Seen through Ergonomic-tinted Spectacles

There are many different types of clerical jobs in which VDTs are used. But from the point of view of ergonomics it is possible to distinguish five kinds of jobs, characterized by predominant modes of interaction with the VDT:

1. Data-entry work
2. Data acquisition
3. Conversational or interactive communication
4. Word processing
5. Computer-aided design (CAD) or computer-aided manufacturing (CAM).

*Data entry*

Data-entry work is characterized by a more or less continuous input of information through the keyboard into the computer. The operator's gaze is mainly directed towards the source documents, looking only occasionally at the keyboard or the display for periodical control of progress. The eyes focus on the text of the source document, but in some cases the display remains in the visual field. Operators mostly use only the right hand to operate the keyboard, while the left hand handles the source document. Constrained postures are frequently observed in these jobs. The working speed is usually very high and 8000–12 000 strokes per hour is not exceptional. Data-entry work is repetitive and often montonous.

*Data acquisition*

Data acquisition or date enquiry involves calling up information from the computer and reading it from the display. Sometimes reading is associated with a search for some specific information. The attention is directed to the screen, sometimes also to the keyboard and source documents. A typical example of data acquisition is the job of telephone information operators. Starr (192) studied a large group of directory enquiry operators who retrieved directory listings on VDTs. These operators spent nearly all their working hours looking at the screen.

*Conversational tasks*    Conversational or interactive jobs involve both data entry and data acquisition. The operator enters data, which are usually more complicated than those in data entry activities, through the keyboard into the computer and watches the results appearing on the display after a certain time delay. These jobs are characterized by a dialogue between the computer and the operator who has some opportunity for making decisions. At conversational terminals the gaze of the operator alternates mainly between source documents and the display. Some reports roughly estimate that the view is oriented about 50% to the source documents and 50% to the display. Elias and Cail (51) recorded the viewing times in a conversational job, shown in Figure 4. The keyboard is operated with both hands and the speed of strokes is low. Airline reservation or airline space control as well as many occupations in banking are typical examples of conversational tasks.

*Word processing*    Word processing comprises text entry, text recall, controlling text for errors, keying in corrections and designing the layout. Secretarial tasks in document preparation and similar operations as well as formatting, proofreading and editing are frequent applications of word processing. The keyboard is used as in normal typing and the screen is watched for a large part of the working time.

*CAD and CAM*    Computer-aided design (CAD) and computer-aided manufacturing (CAM) are techniques using computers for engineering purposes. Predominant applications include mechanical design, printed circuit-board design and electrical schematics. These tasks were formerly carried out by technical draughtsmen using drawing boards. At the design terminal the engineer develops a product in detail, monitoring his work on a graphic video display. The basic workstation elements are the graphic display, a digitizer command tablet with a pen or a 'mouse' as a cursor control and an alphanumeric keyboard used as a data or command entry device. Activities at CAD or CAM workstations could also be considered as conversational tasks. The worker's control over the job task is great; CAD activities are considered interesting and challenging.

*Visual scanning*    VDT jobs have seldom been systematically analysed and compared with non-VDT jobs. An interesting analysis was conducted by Elias and Cail (51) who studied the visual scanning at a data entry and a conversational task with the NAC eye recorder equipment. These results are shown in Figure 4.

In the data entry task the operators glance at the screen from time to time, while their eyes are chiefly directed towards

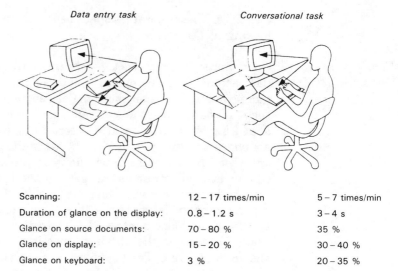

|  | Data entry task | Conversational task |
|---|---|---|
| Scanning: | 12 – 17 times/min | 5 – 7 times/min |
| Duration of glance on the display: | 0.8 – 1.2 s | 3 – 4 s |
| Glance on source documents: | 70 – 80 % | 35 % |
| Glance on display: | 15 – 20 % | 30 – 40 % |
| Glance on keyboard: | 3 % | 20 – 35 % |

**Figure 4 Frequency of scanning and duration of the gaze fixed on the screen in a data entry and in a conversational task.**
*The percentages refer to the approximate duration related to the working time. According to Elias and Cail (51).*

the source documents. In the conversational job the operators rarely change their direction of sight and their eyes are focused on the display for much longer periods. In a group engaged in data acquisition the time spent looking at the display came to a mean duration of 80% of the working time with glances lasting up to 135 s.

*Work-sampling in CAD*

van der Heiden *et al.* (84) carried out a work-sampling study on CAD workstations to determine the relative use of the keyboard, digitizer tablet and screen. Thirty-eight workstations with different engineering tasks (mechanical design, printed circuit-board design and electrical schematics) were involved. The results are shown in Table 1.

The operator's glance was most frequently directed at the screen: between 48 and 68% of the working time at the CAD terminal. This means that an operator observed the graphic display for between $2\frac{1}{2}$ and $3\frac{1}{2}$ hours per day. It follows that the time spent looking at the screen in a CAD task is about the same as that in a conversational VDT task.

Operating the keyboard required 14–24% of the working time. These figures are lower than those observed at other terminals because frequently used command strings were integrated in the tablet menu and activated by the digitizing pen. The use of two input media, keyboard and tablet, gives rise to interference problems. Since the tablet was the primary input medium, the keyboard was usually placed next to it, either to the left or to the right of the tablet, and the normally

Table 1 Results of a work-sampling study at 38 CAD workstations in three different engineering departments.
*According to van der Heiden et al. (84).*

| Activity | Percentage of work time when working interactively on CAD | | |
|---|---|---|---|
| | Dept A Mechanical design (1681)[*] | Dept B Printed wiring-board design (1670)[*] | Dept C Electrical schematics (1752)[*] |
| Watching the screen (incl. manipulation) | 52 | 68 | 48 |
| Operating keyboard | 14 | 14 | 24 |
| Operating tablet | 48 | 43 | 26 |
| Hard-copy manipulation | <1 | <1 | 1 |
| Document manipulation | 14 | 9 | 16 |

[*]Number of observations.

practised two-handed keying induced a twisted position of the trunk. In cases of extensive keyboard use many operators placed it on top of the tablet, causing an unnatural position of hands and wrists while keying. It was concluded that a flat keyboard which can easily be put on and taken off the tablet would reduce constrained postures of hands.

# 3. Physical Characteristics of VDTs

*The CRT*  The basic element of most VDTs is the cathode ray tube (CRT). This large vacuum tube has the screen face at the front, with the inner front surface coated with phosphor compounds and a thickness of 10–15mm, with a high concentration of lead in order to absorb X-rays.

The cathode in the CRT is a gun that emits a narrow beam of electrons at the phosphor-coated surface of the screen. When the electron beam hits the phosphor layer, visible light is generated. Figure 5 shows a diagram of a CRT.

*Raster scan CRT*  The most common types of display use the raster scan CRT, in which the electron beam is swept horizontally across the phosphor, moved down an increment and swept across the phosphor again. After the bottom line has been swept the beam returns to the top and the process starts again.

*Scan lines*  The image quality of a CRT is related to the number of horizontal raster scan lines. If there are too few lines, vertical

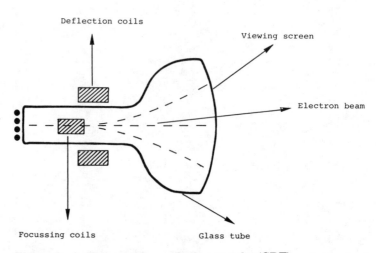

**Figure 5  A diagram of a cathode ray tube (CRT).**

character strokes will appear to be made of distinguishable spots of light rather than continuous strokes.

If the scan line spacing is equal to the spot size of the scanning beam, the spots composing the characters will partially overlap, producing almost stroke-like characters. The more scan lines are used to form a character the better the legibility will be.

Most displays with a 525-line raster present visible spaces between raster lines and cause dot visibility. A well-designed display consisting of 729 or 1029 lines is likely to have raster lines which are barely visible. Beamon and Snyder (12) have shown that visible raster structure is detrimental to legibility. It is therefore concluded that a visible raster structure should be avoided.

*Character formation*

The most common techniques of generating characters on the screen are the dot matrix method or, occasionally, the continuous stroke method. The dot matrix procedure has proved superior in legibility and is preferred by operators.

*Dot matrix*

Here the symbols visible on the screen are composed of dots produced by turning the beam on and off as it sweeps across the phosphor. An ideal grid of dots covers the entire surface of the screen. Each character or number is generated by a matrix of dots, usually $7 \times 9$ or $5 \times 7$ for capital letters, as shown in Figure 6.

If the spaces between dots are greater than the dot diameter the matrix structure of the individual characters will become visible. Snyder and Maddox (187) showed that an increase in the spaces between dots leads to a prolonged reading time. The more a dot matrix character resembles a stroke character the more readable is the text. These authors also compared different dot matrix fonts: a $5 \times 7$ dot matrix was less legible than a $7 \times 9$ matrix, which, in turn, was less legible than a $9 \times 11$ matrix font.

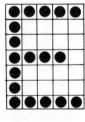
5 × 7

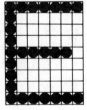
7 × 9

**Figure 6 The character image with the dot matrix system.**

*Refresh rate*

The phosphor on the screen glows for only a fraction of a second after it has been struck by the electron beam, so to produce the illusion of continuous character luminance, the screen must be refreshed over and over again. If the phosphor is refreshed frequently enough, the eye cannot detect the luminance oscillations of the characters. The principle is similar to that of a fluorescent lighting tube, which actually flashes 100 or 120 times each second, too fast to be perceived under most circumstances. The flashing of VDT screens is called flicker. Human sensitivity to flicker depends mainly on the size and brightness of the target image and its location in the visual field. The perception of flicker increases with size and brightness of the target and is likely to appear stronger if the oscillation area is outside the focusing part of the eye. That is why flicker is often only detected if the operator looks at the border or at something outside the screen.

The problems of flicker will be discussed at greater length in the context of measuring the degree of oscillation of character luminance.

*Persistence of phosphor*

Another important factor determining the perception of luminance oscillation on a VDT screen is the persistence of the phosphor applied, that is how long the phosphor remains illuminated after its excitation by the electron beam. The persistence is characterized by the decay time of the phosphor, that is the time the brightness takes to fall down to 10% or 1% of the peak luminance.

If the phosphor decay time is too slow, a 'smearing' of the image may occur. This is also called the 'ghost image', which may appear when scrolling procedures are used. If the phosphor decay time is too fast, characters may appear to flicker.

Whether or not a screen will produce visible flicker is therefore related to both of these physical factors, the persistence of the phosphor and the refresh rate. If visible flicker is to be avoided, a phosphor that glows only briefly will have to be refreshed more frequently than one that has a long decay time. Figure 7 shows the oscillating luminances of two different VDT makes, one with a slow and the other with a fast phosphor.

VDT 'A' has a slow decay time and the luminance does not come to zero between two stimulations of the electron beam. VDT 'G' has a very short decay time and the luminance falls down to zero after a few milliseconds. The result is obvious: in order to get a mean luminance of characters of 40 $cd/m^2$ with VDT 'A' the flash needs a peak brightness of about 75 $cd/m^2$; to obtain about the same mean luminance of characters with VDT 'G', a peak brightness of more than 700 $cd/m^2$ is necessary. It is a fact that VDT 'G' with the fast phosphor is much more liable to flicker than VDT 'A'.

# Physical characteristics of VDTs

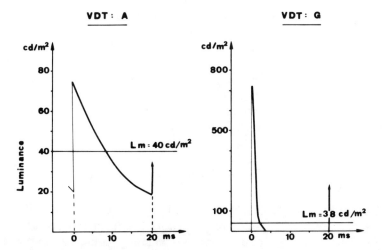

**Figure 7** The oscillograms of character luminances of two VDT makes with different phosphors.
*VDT 'A' (on the left) has a slow phosphor, VDT 'G' (on the right) has a fast phosphor decay time. To get a mean character luminance of 40 cd/m² VDT 'A' needs peak brightness of about 75 cd/m², whereas more than 700 cd/m² are needed with VDT 'G' to get about the same mean character luminance.*
*ms = milliseconds.*

**Table 2 Characteristics of commonly available phosphors for VDTs.** *According to the IBM brochure "Human Factors of workstations with visual displays" (96).*

| Phosphor type | Colour | Decay times (milliseconds) | |
|---|---|---|---|
| | | to 10% brightness | to 1% brightness |
| P1    | green  | 24   | 50   |
| P4    | white  | 0.15 | 0.9  |
| P20   | yellow | 6.5  | 16   |
| P22 G | green  | 6    | 16   |
| P22 B | blue   | 4.8  | 17   |
| P22 R | red    | 1.5  | 23   |
| P31   | green  | 7    | 19   |
| P38   | orange | 1100 | 6000 |
| P39   | green  | 400  | 1300 |
| P45   | white  | 1.5  | 5.2  |

The VDTs on the market today use various phosphors with a wide range of different decay times. Table 2 shows the figures for 10 different phosphors.

*Reversed video*     Most VDTs today have luminous characters on a very slightly illuminated background. They have a so-called positive presentation of characters. Recently a few manufacturers have started offering VDTs with dark characters on a bright background. These are referred to as VDTs with reversed or negative presentation of characters. In reversed VDTs the signal that turns the electron beam on and off is simply reversed, so that a spot that would normally be illuminated is left dark, and spots that would normally be left dark are made to glow. The entire screen is therefore bright except for where the characters appear. This procedure imitates conventional dark letters printed on a light, paper, background. It is said that the reversed presentation reduces the effects of bright reflection on the glass surface but increases the risk of visible flicker. These phenomena will be discussed in detail later.

*Character colour*  Each type of phosphor compound emits light of a characteristic colour. Table 2 shows the colours, related to different phosphors. A certain coloured appearance can also be generated by colour filters which are placed between the phosphor layer and the viewer. With the same procedure it is also possible to give a certain coloured aspect to the background.

The eye is more sensitive to the central part of the visible light spectrum which appears as a yellow-green colour and seems to be brighter than other colours. Many VDT makes have yellow-green characters.

*Chromatic aberration*  Colours may cause problems for the eye's accommodation mechanism through chromatic aberration, resulting from the different refrangibility of various colours. For example, red colours are focused behind the retina, blue colours in front of the retina, while yellow-green is focused exactly on the retina. However, Krueger (115) has shown that the colours used in VDTs are not associated with a noticeable chromatic aberration.

*Preferred colours*  A great number of operators prefer green colours, although they cannot give rational arguments for this choice. It is possible that green characters are more easily distinguished if disturbing reflections appear on the glass surface of the screen. In a Bell Telephone Laboratories publication (13) it is reported that after prolonged viewing of green characters operators tend to see other surfaces, such as white walls, tinged with pink. This phenomenon is called coloured afterimage; it can occur in many situations when the eye is exposed to a strong colour for a long period.

Most VDT ergonomists agree that the colours of characters are not critical, but certain colours at the extremes of the spectrum, such as red, violet and blue, should be avoided.

A few VDT makes have an amber screen background and characters of an illuminated yellow phosphor; many operators like this combination.

Several displays today use more than one colour. Different colours can set off different parts of the text; they function as codes and make some identification processes easier.

It is generally concluded that there is no scientific reason to recommend one colour more than others. The colour of characters remains chiefly a matter of personal preference.

# 4. Vision

## 4.1. The visual system

*Visual perception*

The eyes, acting as receptor organs, pick up energy from the outside world in the form of light waves and convert these into a form of energy that is meaningful to a living organism, i.e., into bioelectric nerve impulses. It is only through the integration of sensory impulses by the brain that we have visual perception. If the afferent sensory nerves, linking the eyes with the brain, are cut, we become blind. Perception in itself does not give a precise photocopy of the world outside: our impressions are a subjective modification of what is perceived. Thus:

A particular colour seems darker when it is seen against a bright background than when it is seen against a darker background.

A straight line appears distorted against a background of curved or radiating lines.

A steady sensation over a long period becomes gradually weaker and can finally vanish.

People differ in the intensity with which they react to visual information: an image can leave one subject indifferent while it may create great emotion in another subject.

*Control mechanisms*

The successive stages of seeing can be simplified as follows: light rays from an object pass through the pupil aperture and converge on the retina. Here the light energy is converted into the bioelectric energy of a nerve stimulus which then passes as a nervous impulse along the fibres of the optic nerve to the brain. At a first series of intermediate nerve cells — called neurons — new impulses are generated which branch off to the centres which control the eyes, regulating the width of the pupils, the curvature of the lens and the movements of the eyeball. These control mechanisms keep the eyes continuously directed at the object, and this is automatic, i.e., not under conscious control. At the same time the original sensory impulses travel further into the brain, and after various filtering processes end up in the cerebral cortex, the seat of consciousness. Here all the signals coming from the eye are

# Vision

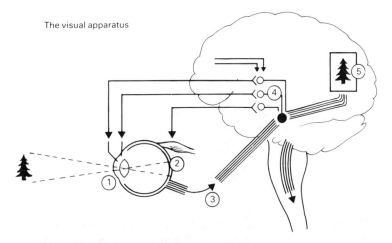

**Figure 8 Diagram of the visual system.**
*1 = cornea and lens. 2 = light received on the retina. 3 = transmission of optic signals along the optic nerve to the brain, 4 = neurons controlling the optic mechanisms of the eyes. 5 = visual perception of the external world in the conscious sphere of the brain.*

integrated into a picture of the external world. Here, too, arise new impulses which are responsible for coherent thought, decisions, feelings and reactions. These processes of the visual system are shown diagrammatically in Figure 8.

In reality, the essential processes of vision are nervous functions of the brain; the eye is merely a receptor organ for light rays. The complete visual system controls about 90% of all our activities in everyday life. It is even more important in a great many jobs in modern offices. If the numerous nervous functions that are under stress during seeing are considered, it is not surprising that the eyes are sometimes an important source of fatigue.

*The eye*

The principal parts of the human eye are shown in Figure 9. The eye has many elements in common with a photographic camera: the retina corresponds to the light-sensitive film,

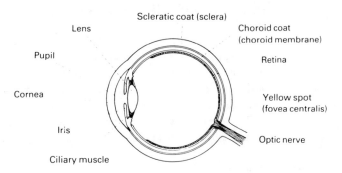

**Figure 9 Diagram of the eye in longitudinal section.**

whereas the transparent cornea, the lens and the pupil with its variable aperture are similar to the optics of the camera. Cornea and lens together refract the incoming rays of light and bring them to a focus on the retina, mainly in the fovea centralis.

*The retina*   The actual receptor organs are the visual cells embedded in the retina, consisting of 'cones' for daylight vision and specially sensitive 'rods' for vision in dim light. The visual cells convert light energy by photochemical reactions into nervous impulses, which are then transmitted along the fibres of the optic nerve.

*The fovea*   The human eye contains about 130 million rods and 7 million cones, each of which is approximately 0.01 mm long and 0.001 mm thick. On the posterior surface of the eye, a few degrees away from the optical axis, is the retinal pit, or *fovea centralis*, characterized by a thinner covering than the surrounding area. The thin covering allows the light rays to pass directly to the visual cells, which, in the fovea, consist entirely of cones, here at their maximum density of about 10 000/mm$^2$. Each foveal cone has its own fibre connecting it to the optic nerve. For these reasons the fovea has the highest resolving power of any part of the retina, up to about 12 seconds of arc. Since vision is most acute in the area of the fovea, it is instinctive to look at an object closely by moving the eye until the image falls upon this area of the retina, which is called the area of central vision. Any object or sign that is to be seen clearly must be brought to this part of the retina, which covers a visual angle of only 1°.

*Cones and rods*   Outside the foveal area the cones are considerably fewer, and one nerve fibre serves several cones and rods. Here the rods are distinctly more abundant than cones, and they become more numerous as the angle from the fovea increases, whereas the number of cones declines. Although rods are more sensitive to light than cones they do not detect such fine differences of either shape or colour. The rods are the more important light-detecting organs in poor visibility and at night.

*The sharp picture*   To summarize, only objects focused upon the fovea are seen clearly, while other images become progressively less distinct and blurred as distance from the fovea increases. Normally the eye moves about rapidly, so that each part of the visual field falls on the fovea in turn, allowing the brain to build up a sharp picture of the whole surrounding.

*The visual field*   The visual field is that part of one's surroundings that is taken in by the eyes when both eyes and head are held still. Only

# Vision

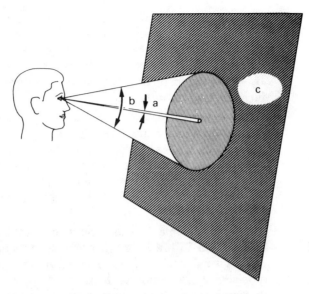

**Figure 10 Diagram of the visual field.**
a = *zone of sharp vision; angle of view of 1°.* b = *middle field; vision unsharp; angle of view from 1° to 40°.* c = *outer field, only movements perceptible; angle of view from 41° to approximately 70°.*

objects or signs within a small cone of 1° apex are focused sharply. Outside this zone objects become progressively more blurred and indistinct. If the eyes are kept still when reading only a few letters can be focused. More details about the physiology of reading will be discussed later in this chapter.

As shown in Figure 10, the visual field can be divided up as follows:

(a) area of distinct vision: viewing angle 1°
(b) middle field: viewing angle 40°
(c) outer field: viewing angle 40–70°

Objects in the middle field are not seen clearly, but strong contrasts and movements are noticed: alertness is maintained by quickly shifting the gaze from one object to another. The outer field is bounded by the forehead, nose and cheeks; objects in this area are hardly noticed unless they move.

## 4.2. Accommodation

*Accommodation means the ability of the eye to bring into 'sharp focus' objects at varying distances* from infinity down to the nearest point of distinct vision, called the 'near point'. If we hold up a finger in front of the eye, the finger can be focused sharply, leaving the background blurred, or the background can be focused sharply, leaving the finger indistinct. This demonstrates the phenomenon of accommodation.

An object is seen clearly only when refraction through the cornea and lens produces a tiny but sharp image on the retina, the three components forming an optical system. Focusing on near objects is achieved by changing the curvature of the lens, by contraction of *the muscles of accommodation*, called *the circular ciliary muscles*.

*Distant objects*

When the ciliary muscles are relaxed, the refraction of the cornea and lens is such that parallel rays from distant objects are focused on to the retina. Therefore, when attention is allowed to wander over distant objects, the eyes are focused on 'infinity' and the ciliary muscles remain relaxed.

*Resting accommodation*

For a long time it was assumed that the accommodation focused on infinity was also the resting position of the eye. But several studies have revealed that in the dark the resting position corresponds to focusing distances lying somewhere between the near point and infinity. Krueger and Hessen (116) determined for students in a resting position a mean focusing distance of 80 cm. This distance seems to move gradually towards infinity with increasing age.

*Near vision*

Without accommodation the image of an object nearer to the eye would fall behind the retina, which would receive a blurred impression. To avoid this the ciliary muscle increases the curvature of the lens so that the image is brought back into the plane of the retina. In near vision the lens is continuously adapting the focal length in such a way that a sharp image is projected on the retina. To maintain focus on a near object the ciliary muscle must continuously exert a contracting force.

The accommodated lens is in constant motion. When viewing a target the lens will oscillate in a certain range at a rate of about 4 times per second. Even when reading a book the lens remains quite active. It seems that these movements of the lens as well as the perception of blur are important for the automatic regulation of accommodation. The key to comfortable viewing is accommodation; it means that the image is well focused on the retina.

After viewing a near object for some time the lens may not immediately return to its relaxed position. This condition, referred to as 'temporary myopia', may remain for several minutes.

*The near point*

As already mentioned, the shortest distance at which an object can be brought into sharp focus is called the near point and the furthest away is the far point. The nearer the object is, the greater is the load on the ciliary muscle to bring it into focus and keep it in focus. The near point is a measure of the power of the ciliary muscle and of the elasticity of the lens. It moves further away as the ciliary muscle becomes tired after a long spell of close work. Many experiments have shown that prolonged reading under unsuitable conditions is associated with increased figures of the near point, a phenomenon considered as a symptom of visual fatigue.

*Age and accommodation*

Age has a profound effect on powers of accommodation, because the lens gradually loses its elasticity. As a result the near point gradually recedes, whereas the far point usually remains unchanged or becomes slightly shorter.

The average distance of the near point at various ages is reported in Table 3.

Table 3 Average near point distance at different ages.

| Age (years) | Near point (cm) |
|---|---|
| 16 | 8 |
| 32 | 12 |
| 44 | 25 |
| 50 | 50 |
| 60 | 100 |

*Presbyopia*

When the near point has receded beyond 25 cm close vision becomes gradually more strenuous, a condition called presbyopia. It is caused by the loss of elasticity of the lens due to age. This prevents the lens from changing its curvature. The way to correct for presbyopia is to wear glasses.

Presbyopia is a frequent reason for visual discomfort while doing close work. It is due to the increased static muscle strength which is needed to compensate for the loss of lens elasticity. This additional muscular activity might be one of the reasons for visual fatigue. It is said that no more than two-thirds of the available accommodation power should be used to maintain a comfortable degree of focus.

*Speed and accuracy of accommodation*

The level of illumination is a critical factor in accommodation. When the lighting is poor the far point moves nearer, and the near point recedes, while both speed and precision of accommodation are reduced as well as luminance contrast and sharpness of printed texts; the sharper the object or the character stands out against its background the quicker and more precise the accommodation.

The speed as well as the precision of accommodation decreases with age. According to Krueger and Hessen (116) these two functions show a marked decrease about from the age of 40 on.

## 4.3. The aperture of the pupil

*The 'diaphragm' of the eye*

Two different muscles control pupil aperture: one constricting and the other widening the pupil size. This part of the eye is called the iris. Its function can be compared to the one of the diaphragm of a camera which is used to avoid under- and over-exposure. The pupil aperture is under reflex control to adapt the amount of light to the needs of the retina. When light levels increase the iris contracts and the pupil size is reduced. When light levels are decreased the iris opens, making the pupil larger. In any given lighting condition the pupil is in a resting position as soon as the pupil size has stabilized. Even in this state, however, the pupil is in constant motion, much like the accommodated lens of the eye.

*Speed of pupil reaction*

The adjustment of the aperture of the pupil takes a measurable time which may vary from a few tenths of a second to several seconds. Fry and King (59) demonstrated that when stimuli producing a significant change in pupil size are presented at a slightly higher rate, by about 3 Hz, than the pupil can respond to, the pupil reaction is dampened and discomfort is produced. In fact, if the level of lighting changes frequently and strongly, there is a danger of over-exposure of the retina, since the reaction time of the pupil is comparatively slow.

*Brightness and pupil size*

Pupil size reflects to a large extent the brightness of the visual field. It seems that the central vision is of greater importance for the regulation of the pupil size than the outer areas of the retina. During daylight the aperture may have a diameter of 3–5 mm, increasing at night to 8 and more mm. Experiments revealed for extreme conditions a range in pupil area from 5 $mm^2$ up to 40 $mm^2$.

*Other regulating factors*

The aperture of the pupil is also affected by two other factors:
1. The pupil contracts when near objects are focused and opens when the lens is relaxed.
2. The pupil reacts to emotional states, dilating under strong emotions such as alarm, joy, pain or intense mental concentration. The pupil contracts under fatigue and sleepiness.

Under normal conditions, however, the general level of lighting is the dominant regulating factor of pupil size.

*Pupil size and acuity*

When the pupil becomes smaller, the refractive errors of the lens are reduced and this improves visual acuity. One of the reasons that higher levels of lighting increase visual acuity is the narrowing effect of light on the pupil size. Here too, it is possible to make a comparison with the camera: a small aper-

# Vision

ture of the diaphragm will increase depth of field and generate a sharper image.

## 4.4. The adaptation of the retina

If we look at the headlights of a car at night we are dazzled, but the same headlights do not dazzle in daylight. If we walk from daylight into a darkened cinema where the film has already started we can see very little at first, but after about 10–20 minutes the surroundings of the theatre gradually become visible. These are everyday examples of how the sensitivity of the retina is continuously adapted to the prevailing light conditions. In fact, this sensitivity is many times higher in darkness than in daylight.

The process is called adaptation and comes about through photochemical and nervous regulation of the retina. Thanks to this facility we can see nearly as well in moonlight as in the brightest sunlight, even though the level of illumination has decreased more than 100 000 times.

*Adaptation to darkness*

Adaptation to darkness or to brightness takes a comparatively long time. Darkness adaptation is very quick in the first five minutes, afterwards it becomes progressively slower; 80% adaptation takes about 25 minutes and full adaptation as much as one hour. Hence sufficient time must always be allowed for darkness adaptation, at least 25–30 minutes for good night vision.

*Adaptation to light*

Light adaptation is quicker than darkness adaptation. The sensitivity of the retina can be reduced by several powers of ten in a few tenths of a second. Yet light adaptation, too, continues over a measurable time of the order of several minutes.

The abrupt reduction of sensitivity during light adaptation

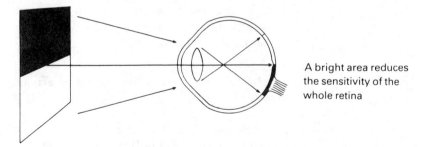

A bright area reduces the sensitivity of the whole retina

**Figure 11  Effects of bright and dark surfaces on the retina.**
*A bright patch reduces the sensitivity of the entire retina and thereby reduces the visual acuity in the fovea. This form of disturbance is called relative glare.*

involves the entire retina. Whenever the image of a bright surface (a window, a light source or a bright reflection) falls on to any part of the retina, sensitivity is reduced all over, including the fovea centralis. This phenomenon, most important for precision work or for reading tasks, is illustrated in Figure 11.

*Partial adaptation*

In other words: if the visual field contains a dark or a bright area, adaptation will occur in the corresponding part of the retina. This adaptation appears in one part of the retina and is called local or partial adaptation. But, as mentioned before, this partial adaptation spreads over the whole retina, including the fovea. Such partial adaptation will therefore change the sensitivity of the retina and affect vision.

Furthermore, adaptation of one eye has some corresponding effect on the other, a fact that may be significant at working places where only one eye is employed.

*Ergonomic principles*

Two general ergonomic principles can be deduced from this knowledge:

1. *To avoid dazzle effects, all important surfaces within the visual field should be of the same order of brightness.*
2. *The general level of illumination should not fluctuate rapidly because pupil reaction as well as retinal adaptation is a relatively slow process.*

*Glare*

Physiologically speaking, *glare is a gross overloading of the adaptation processes of the eye, brought about by overexposure of the retina to light*. Three types of glare may be distinguished:

1. *Relative glare*, caused by excessive brightness contrasts between different parts of the visual field.
2. *Absolute glare*, when a source of light is so bright (e.g., the sun) that the eye cannot possibly adapt to it.
3. *Adaptive glare*, a temporary effect during the period of light adaptation; e.g., on coming out of a dark room into bright daylight outside. This phenomenon is also called 'transient adaptation'.

*Practical hints*

In this context the following hints are important for the layout of work places:

1. The effect of relative glare is greater the nearer the source of dazzle is to the optical axes, and the larger its area.
2. A bright light above the line of sight is less dazzling than one below, or to either side.
3. The risk of dazzle is greater in a dim room since the retina is then at its most sensitive.

## 4.5. Eye movements

*Tremor of eye muscles*

The eyeball has several external muscles which direct the eye to the point of interest. It continuously makes small movements which keep the retinal image in constant, slight, motion. Without that continuous tremor the perceived image would fade away. This is like placing your hand lightly on a rough surface and feeling the roughness only as long as the fingers move back and forth.

In general, eye movements are very precise and fast. An eye movement of 10° may be accomplished in about 40 milliseconds.

*Vergence movements*

For good vision the so-called vergence movements or convergence are of special importance. Binocular vision requires the optical axes of the two eyes to meet at the object being looked at, so that the image falls onto the corresponding part of the retina in each eye. When viewing an object relatively near, the visual axes are turned slightly inwards in order to intersect at the distance of the object being viewed.

If the gaze is shifted to a second object, further away, the angles of the two eyes must be opened until the optical axes again cross at the object. This movement is brought about by activity of the outer eye muscles; it is a very delicate adjustment upon which distance perception depends. This specific sensitivity is gradually developed in infancy until we have finally learned by experience to estimate distance mainly from angular convergence of our two eyes. In monocular vision distances must be guessed from the apparent size of objects, from foreshortening by perspective and from other visual experiences.

*The incredible number of eye movements*

The number of eye movements required when reading a book may be as high as 10 000 coordinated eye movements per hour (96). Walking over a rocky track in the mountains demands even more from the eye muscles. When the head is in motion, as in walking, the external eye muscles are in constant activity to adjust the position of the eyes in order to maintain steady fixation points. That is why objects viewed by an observer, even when walking or sitting in a car, appear stable.

If the coordination of external eye muscles is disturbed the phenomenon of double images will appear. This can be easily demonstrated by slightly touching one eyeball with a finger. In case of excessive fatigue transitory double images can cause annoying sensations.

## 4.6. Visual capacities

The various functions of the eye are not usually pushed to the limits of their performance in everyday life, but it may sometimes occur in industry or under modern traffic conditions. Furthermore, visual performances are often used in laboratory experiments to evaluate the effects of various variables such as lighting or other viewing conditions. The most important visual capacities are:

Visual acuity
Contrast sensitivity
Speed of perception

*Visual acuity*

Visual acuity is the ability to perceive two lines or points with minimal intervals as distinct, or to determine the form and shape of signs and discern the finest details of an object. By and large, visual acuity is the resolving capacity of the eye. The ability to resolve a one minute of arc wide separation between two signs is often considered as 'normal' acuity. In this case the minimum distance between two points in the image on the retina is $5 \times 10^{-6}$ m. However, under adequate lighting conditions a person with good vision should be able to resolve an interval of about half that distance.

*Influences on visual acuity*

Visual acuity is related to illumination and to the nature of observed objects or signs as follows:

Visual acuity increases with the level of illumination, reaching a maximum at illumination levels above 1000 lx. The increase is about 150%.

Visual acuity increases with the contrast between the test symbol and its immediate background, and with the sharpness of signs or characters.

Visual acuity is greater for dark symbols on a bright background than for the reverse. (Bright background decreases the pupil size and reduces refractive errors.)

Visual acuity decreases with age, which is shown in Figure 12.

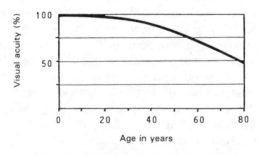

Figure 12 Decrease of visual acuity with age.
According to Krueger and Müller-Limmroth (114).

*Contrast sensitivity*

Sensitivity to contrast is the ability of the eye to perceive the smallest difference in luminance, and thus to appreciate

Vision

niceties of shading and the slightest nuances of brightness, all of which may be decisive for the perception of shape and form. Contrast sensitivity is more important in everyday life than visual acuity, and this also applies for many jobs of inspection and product control. In order to measure contrast sensitivity a four-minute diameter disc is frequently used as a target. The threshold of perceiving a luminance difference between target and background is used to determine contrast sensitivity.

*Influences on contrast sensitivity*

Contrast sensitivity is subject to the following rules:

1. It is greater for large areas than for small ones.
2. It is greater when boundaries are sharp and decreases when the change is gradual or indefinite.
3. It increases with the luminance of the surroundings and is greatest within the range of 0.01 to more than 1000 cd/m$^2$ (96).†
4. It obeys the Weber–Fechner Law;‡ within the mentioned range a contrast equal to about 2% of the surrounding luminance can be observed.
5. It is greater when the outer parts of the visual field are darker than the centre, and weaker in reverse contrast.

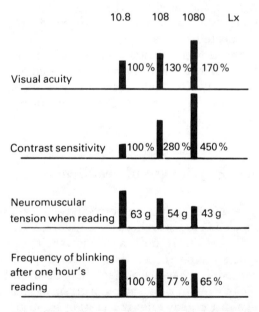

Figure 13 Effects of three increasing illumination levels on visual acuity, contrast sensitivity, nervous strain and eye blinking rate. *According to Luckiesh and Moss (134).*

---

† For a definition of luminance and illumination see Chapter 5, "Ergonomic principles of lighting in offices".

‡ The Weber–Fechner Law states that the physiological sensation produced by a stimulus is proportional to the logarithm of the stimulus.

Figure 13 shows results of experiments carried out in 1937 by Luckiesh and Moss (134). It appears that raising the illumination level from approximately 10 lx to 1000 lx increases visual acuity from 100 to 170% and contrast sensitivity to 450%. At the same time the investigators recorded a decrease of muscular tension (measured from the continuous pressure of a finger on a key) and rate of blinking of the eyelids. This was interpreted as a reduction of nervous tension as a result of better lighting.

*Speed of perception*

The speed of perception is defined as the time interval that elapses between the appearance of a visual signal and its conscious perception in the brain. Speed of perception is commonly measured by the technique of tachistoscopy. In this procedure a series of words is presented to the test subjects for a short time. The minimum display time required for correct perception is measured and used as a parameter. Speed of perception measured with such a procedure is of course mainly a function of neural and mental mechanisms in the brain.

Speed of perception increases with improved lighting as well as with higher luminance contrast between an object (or sign) and its surrounding. That means that lighting, visual acuity, contrast sensitivity and speed of perception are closely connected with each other.

Speed of perception can be vital in transport. We need only think of an aircraft flying at the speed of sound, and how much can happen at that speed during a perception time of 0.2 s, a common figure. But speed of perception is also an important factor in reading.

## 4.7. Physiology of reading

*Saccades*

There is a distinction between reading, the taking in of information, and search, which involves the locating of needed information. In both activities the eyes move along a line in quick jumps rather than smoothly. These jumps are called *saccades*. They are so fast that no useful information can be picked up during their occurrence. Between the jumps the eyes are steady and fix a certain small area which is projected. Only in the fovea and in the adjacent area, the para-fovea, is detailed vision sufficiently accurate for the recognition of normal print.

Three forms of reading saccades are of importance: the rightward reading saccades, the correction saccades and the leftward line saccades.

*Rightward reading saccades* along a line cover in each jump an area of about $8 \pm 4$ letters. Occasionally small leftward

# Vision

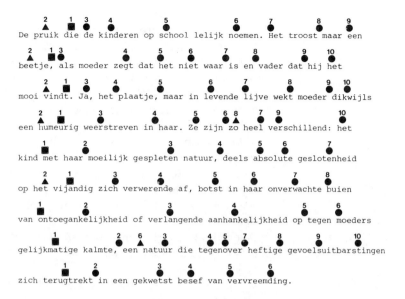

**Figure 14  Saccades and fixations of eyes in the silent reading of a Dutch text.**
*Three different types of saccades are indicated: reading saccades (circles), correction saccades (triangles) and line saccades (squares). Numbers indicate the order of fixations within each line. According to Bouma (18).*

saccades may occur, the so-called *correction saccades*. *Line saccades* start just before the end of a line is reached and jump to the beginning of the next line.

Bouma (18) did a thorough study on eye saccades and eye fixations of reading subjects. Figure 14 shows the succession of eye saccades and eye fixations of a subject reading a Dutch text. All types of saccades may be different for different texts and different subjects.

*Character recognition*

The eye pauses between saccades last mostly between 120 and 300 ms (Bouma (18)). During these pauses characters are recognized in foveal and parafoveal vision. For quick and good recognition it is important that characters are acceptable, identifiable and distinctive (18).

*Acceptability* is the degree to which characters correspond to the 'internal mode' readers have of them. This is the fundamental process of reading.

*Identifiability* requires clear letter details which must be designed clearly.

*Distinctiveness* means that each character has such a specific design that no confusion can occur. The extension of descending letters (such as p and q) and of ascending letters (b or d) can be important for good distinctiveness.

*Visual reading field*

During the eye pause the fovea and the adjacent area pick up visual information from a rather small surface, the so-called visual reading field. In order to code numbers without much redundancy only a few symbols can be picked up in a single glance. For words the visual reading field may be larger, because sufficient word knowledge renders the recognition of the full word possible at the sight of merely a few letters.

When reading a text the eyes make about four fixations per second. In well printed texts the visual reading field can easily be as wide as 20 letters, about 8 to the left of fixation and 12 to the right. The visual reading fields overlap, that is to say that words within the visual reading field may appear at least twice.

According to Dubois-Poulsen (48) the following time fractions are about normal:

```
Gaze fixation between saccades: 0.07 – 0.3 s
Rightward reading saccades:     0.03 s
Line saccades:                  0.12 s
Number of saccades per line:    about 6
```

*Line saccades*

Correct line saccades require sufficiently large distances. The lines above and below the reading line will interfere with parafoveal word recognition unless line distances are sufficiently wide. If they are too narrow the visual reading field becomes restricted so that less information can be picked up in a single eye pause. *Thus a wide visual reading field calls for sufficient inter-line distance.* According to Bouma (16), the visual reading area around fixation which is free from interference by the two adjacent lines of print, decreases with a reduction in the interline distance. If the reading field covers 15 letters the interline distance must be equal to about 5 times the height of the lower case characters; if the reading field is restricted to 7 letters the interline distance must still be equal to two lower case characters. The same author recommends a minimum admissible interline distance of about 1/30 of the line length (this text has an interline distance of 1/24 of the line length). As a consequence, interline distance should increase with line length. For VDTs it seems advantageous to use screens vertically oriented, since such a screen design would require shorter lines and smaller interline distances.

*Contrast and colour*

According to Timmers (197), parafoveal word recognition is critically dependent on character contrast. The lower the contrast, the narrower the visual reading field and the lower, therefore, the readability. Similar effects were observed with coloured letters. Engel (53) showed that coloured letters and digits can only be read when quite close to the fixation, although colour itself may well be discernible far away from the fixation. *This indicates that colour is a useful aid for visual search but actual reading takes place in a restricted visual reading field.*

If a reader is familiar with the significance of colours, then colours will help to locate the required information quickly, but the recognition of a word or symbol itself depends on the legibility of characters and not on their colour.

# 5. Ergonomic Principles of Lighting in Offices

## 5.1. Light measurement and light sources

In order to understand what follows later it will be useful to define two of the many terms employed in the study of lighting, illumination and luminance.

*Illumination*  *Illumination is the measure of the light falling on a surface.* The light may come from the sun, lamps in a room or any other bright source. The unit of measurement is the lux, defined as:

1 lux (lx) = 1 lumen per square metre (lm/m$^2$), a lumen being the unit of luminous flux.

A previously used unit in the English speaking world was the footcandle (fc). 1 lux is approximately 0.1 footcandle (0.0929 fc).

The human eye responds to a very wide range of illumination levels, from a few lux in a darkened room to approximately 100 000 lx outside in the midday sun. Illumination levels in the open vary between 2000 and 100 000 lx during the day, whereas at night artificial light of 50–500 lx is normal.

*Luminance*  Luminance is the measure of the brightness of a surface; the perception of brightness of a surface is proportional to its luminance. *Luminance is therefore a measure of light coming from a surface.* Since it is a function of the light that is emitted or reflected from the surfaces of walls, furniture and other objects, it is greatly affected by the reflective power of the respective surface. The luminance of lamps on the other hand is an exact measure of the light they emit. Bright characters on a dark background in VDTs are also emitting light which can be characterized by its luminance.

*Candles/m$^2$*  In the metric system *luminance is measured in units of candles per square metre* (cd/m$^2$). In the past the standard reference for measuring luminance was actually a wax candle of a certain specification. Today the standard is much more precise, but the terminology stems from the earlier concept.

## Ergonomic principles of lighting in offices

**Millilambert and Footlambert**

In the English speaking world the terms millilambert (mL) and footlambert (ft L) are still used to measure luminance. One millilambert (mL) is the amount of light emitted from a surface at the rate of 0.001 lumen/cm². A footlambert (ft L) is the amount of brightness of an ideally reflecting surface illuminated by one footcandle.

The following equations apply:

1 cd/m² = 0.292 footlambert (ft L)
1 footlambert (ft L) = ca 3.5 cd/m²
1 millilambert (m L) = 3.183 cd/m²
1 footlambert (ft L) = 1.076 mL

Fortunately, in the English speaking world today the cd/m² has gradually become the most frequently used unit to define the luminance of surfaces.

A few examples will illustrate the approximate luminance of some common sources of light in an office with an illumination of 300 lx:

| | |
|---|---|
| Fluorescent lamp (65 watt) | 10 000 cd/m² |
| Window surface | 1000–4000 cd/m² |
| White paper lying on a table | 70–80 cd/m² |
| Table surface | 40–60 cd/m² |
| Bright enclosure of a VDT | 70 cd/m² |
| Dark enclosure of a VDT | 4 cd/m² |
| Screen background | 5–15 cd/m² |

**Reflectance**

If the luminances of various surfaces are compared they can also be expressed as reflectance, which is the ratio between incident and reflected light. *Reflectance is usually expressed as the percentage of reflected to incident light.* The luminance in cd/m² and the illumination in lx are related as follows:

$$\text{reflectance } (\%) = \frac{0.32 \text{ cd/m}^2}{\text{lx}}$$

A simple example is this: if a bright table surface has a reflectance of 70% and the incident light has an illumination figure of 400 lx, the luminance of the table will then be 70% × (0.32 × 400) = 89 cd/m².

**Direct and indirect lighting**

Among the various lighting systems in offices one can distinguish between direct and indirect lighting.

*Directional lighting* sends about 90% of its light towards targets in the form of a cone of light. These light sources cast hard shadows with sharp contrasts between light and shadow. Excessive contrast tends to produce relative glare. Directional lighting systems can be recommended in offices as working lights only where the general illumination is high enough to reduce this contrast. At VDT workstations such lighting is used when the general illumination is insufficient for reading source documents with poor legibility.

*Indirect lighting* throws 90% or more of its light onto the ceiling and walls which reflect it back into the room. This system requires the ceiling and walls to be light-coloured. Indirect lighting generates diffuse light and casts practically no shadows. In a traditional office (without business machines) it can give a high level of illumination with a low risk of glare. In offices with VDTs the bright ceilings and walls can produce reflections on the screens and cause relative glare.

A widely used system in offices is a combination of direct and indirect lighting. The lights have a translucent shade and about 40–50% of the light radiates to the ceiling and walls while the rest is thrown directly downwards. This type of lighting casts only moderate shadows with soft edges. The whole room, including furniture and shelves placed against the walls, are evenly lit.

In offices with VDTs special attention must be paid to lighting; these aspects will be discussed in detail in Section 5.5, 'Appropriate lighting', with relation to screen reflections.

Light sources are mainly of two kinds: electric filament lamps and fluorescent tubes.

*Filament lamps*

The light of filament lamps is relatively rich in red and yellow rays. When used above a workplace they emit heat. Lampshades can reach temperatures of 60°C and more and can cause discomfort and headaches. On the other hand, their warm glow does create a pleasant atmosphere.

*Fluorescent tubes*

Fluorescent lighting is produced by passing electricity through a gas (argon or neon) or through mercury vapour. This procedure converts electricity into light much more efficiently than a heated filament. The inside of the tube is covered with a fluorescent substance which converts the ultraviolet rays of the discharge into visible light, the colour of which can be controlled by the chemical composition of the fluorescent material. Fluorescent tubes have a series of advantages:

*Advantages*

*High output of light and long life.*
*Low luminance, when adequately shielded.*
*Ability to match the light to daylight or at least to a pleasant and slightly coloured light.*

However, fluorescent lighting also has serious drawbacks:

*Drawbacks*

Since they operate from alternating current, fluorescent tubes produce flickering light at a frequency of 100 Hz in Europe and 120 Hz in the USA. *This is above the normal flicker fusion frequency, the so-called critical fusion frequency (CFF) of the human eye*, but it can become noticeable as a stroboscopic effect on moving objects. Furthermore, old or defective tubes develop a slow visible flicker.

*Visible flicker*

*Visible flicker has adverse effects on the eye mainly because of the repetitive overexposure of the retina. Flickering light is extremely annoying and causes visual discomfort.*

It is generally assumed that the luminance oscillation of fluorescent tubes with a rate of 100 Hz and more is above the critical fusion frequency and can therefore not affect the eye. Several studies, though, indicate that the exposure to single fluorescent tubes may have adverse effects on human subjects. Our own studies (65) have shown that working lights with single fluorescent tubes can increase visual fatigue and measurably reduce the performance of fine assembly work. Wiebelitz and Schmitz (209) observed a decrease of the pupil reaction to light when subjects were reading under fluorescent light with flicker frequencies between 25 and 100 Hz. Raising the flicker frequency to 200 Hz or more removed this effect. Recent experiments on cats carried out by Eysel and Burandt (54) revealed that the visual system in the brain responds distinctly to the temporal information present in light from fluorescent tubes driven by 50 or 60 Hz alternating current. This study confirmed earlier findings showing that the critical fusion frequency in the optic tract is above 100 Hz, but cannot be considered to be definite proof of adverse effects of fluorescent light with 100 or 120 Hz invisible flicker.

*Early experiences with fluorescent light*

A general experience of rather anecdotal character must be mentioned here: when fluorescent lighting was first introduced on a large scale in European offices, a series of complaints about irritated eyes and eye strain were reported. On the assumption that the oscillating character of fluorescent light was the cause of visual discomfort, the lighting manufacturers developed phase-shifted equipment which produced almost constant light. Complaints seem to have stopped in offices where phase-shifted fluorescent tubes were installed.

*Subharmonic 50 Hz oscillations?*

A study by Collins (37), carried out in 1957, revealed another interesting aspect of fluorescent tubes. With a number of different models of fluorescent tubes he recorded small 50 cycle per second fluctuations superimposed on the main 100 cycle one. These subharmonic 50 Hz oscillations come from a partial rectifying action in the discharge due to asymmetrical emission of the electrodes. 10% of both old and new tubes were observed to have this effect. Small amounts of subharmonics were found to be perceptible by subjects and the author asssumed that such tubes are sufficiently common to account for complaints which had arisen with fluorescent lighting. Nevertheless, the degree of oscillation in fluorescent tubes that is acceptable or not has not yet been determined and agreed upon.

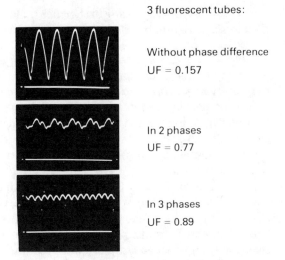

**Figure 15 Recordings of the luminance oscillation of fluorescent tubes.**
*Upper section: one single fluorescent tube without phase shifted equipment. Middle section: two fluorescent tubes with oscillations shifted in two phases. Lower section: three fluorescent tubes with oscillations shifted in three phases. Horizontal lines = zero level of luminance. UF = Uniformity figure, which is the minimum divided by peak luminance.*

*Phase shifted fluorescent tubes*

In Europe it was generally resolved that *offices should never be lit with single fluorescent tubes but always with two or more phase shifted tubes inside one lighting unit.*

Figure 15 shows recording of the luminance oscillation of fluorescent tubes. It illustrates the effects of phase-shifted equipment which generate an almost constant luminance.

It is obvious that appropriate equipment will avoid the disadvantages of fluorescent tubes, so that their undisputed advantages can be fully utilized. The oscillation of character luminance of VDTs will be discussed later in Section 6.4.

## 5.2. Illumination levels

*Physiological requirements*

To achieve visual comfort and good performance the following conditions should be met:

> Suitable level of illumination
> Spatial balance of surface luminances
> Temporal uniformity of lighting
> Avoidance of glare with appropriate lights

50 years ago illumination levels of about 100 lx were generally recommended for offices. Since then the figures have increased steadily and today levels between 500 and 2000 lx are quite common. The general attitude towards lighting has

# Ergonomic principles of lighting in offices

**Table 4 Sources of glare in 15 open-plan offices.**
*Percentage refers to statements of 120 complaining employees. According to Nemecek and Grandjean (155).*

| Sources of glare | mentioned by |
|---|---|
| Windows | 36% |
| Lights | 25% |
| Polished tabletops | 18% |
| Contrasts on desks | 17% |
| Other causes | 4% |

*Drawbacks of too high illumination levels*

been 'the more the better'. But this does not necessarily hold true for offices. A study of 15 open plan offices (155) has shown that a very high level of illumination is often unsuitable in practice. Levels above 1000 lx increase the risk of troublesome reflections, deep shadows and excessive contrasts. In this study 23% of 519 employees reported that they were disturbed by either reflections or glare. The reported sources of glare are given in Table 4.

An interesting observation was the higher incidence of eye troubles in offices with illumination levels above 1000 lx. The results of this comparison between moderate and high illumination levels are presented in Figure 16.

All employees generally preferred illumination levels between 400 and 850 lx, which is in accordance with the results

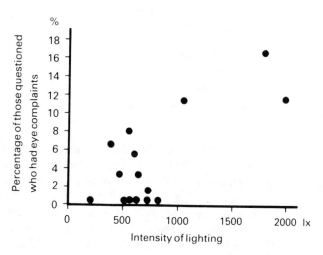

**Figure 16 The incidence of eye complaints in relation to the illumination levels in each of the 15 open-plan offices.**
*519 employees. 100% is the number of employees questioned in each office. The incidence difference between offices with illumination levels of 1000 lx and more and those with 200–800 lx was statistically highly significant (p <0.001). According to Nemecek and Grandjean (155).*

shown in Figure 16. Obviously it would be going too far to interpret these results to the effect that there is a direct causal relationship between illumination level and eye troubles, but there are good reasons to believe that brightly lit open plan offices more often create reflections, deep shadows and relative glare, and that these possibly contribute to the eye troubles recorded. In other words: illumination levels of over 1000 lx do not directly cause visual discomfort; it is their effect on the balance of luminances which is the main source of trouble. These results conflict with several studies carried out in well-lit test rooms with a preferred illumination level of 1000–4000 lx. The carefully designed brightness of the surrounding may account for the contradictory results.

*Recommendations for offices without VDTs*

A comparison of current recommendations will show that the standards of the US Illumination Engineering Society (IES) prescribe higher illumination levels for offices than European guidelines. For office work such as book-keeping the German DIN (44) recommends 500 lx whereas the corresponding figure of the IES (95) is as high as 1600 lx.

*Based on general experience illumination levels between 500 and 700 lx can be recommended for most office jobs which do not use VDTs.*

*Illumination and age*

As already mentioned, several visual functions lose some of their capacity with age, which is why illumination levels must be increased with age if constant visual performances are to be maintained over the years. According to Fortuin (56) the lighting levels required for reading a well-printed book must be multiplied by the following factors:

```
20–25 years old   1.0
40 years          1.17
50 years          1.58
65 years          2.66
```

*Illumination levels for VDT workstations*

The above mentioned general recommendations for illumination levels are not valid for offices with VDT workstations. A VDT operator who is alternately looking at a dark screen and a bright source document is exposed to great luminance contrasts. It will be shown on the following pages that the contrast ratio between screen and source document should not exceed a figure of 1:10, which implies that the illumination level on source documents should be kept low. On the other hand, however, the reading task requires that the source document be well illuminated. This conflicting situation calls for a compromise. Hence it is not surprising that the assessment of the optimum illumination level is a controversial matter.

**Table 5** Illumination levels measured on horizontal planes at 38 CAD workstations.
*Many operators had switched off some of the lights or drawn the curtains and blinds. According to van der Heiden et al. (190).*

| Workstation element | Illumination levels (lx) | |
|---|---|---|
| | Median | 90% range |
| Tablet | 125 | 15–440 |
| Keyboard | 125 | 15–505 |
| Reference table | 118 | 15–500 |

*Preferred illumination levels*

Walking through offices with VDT workstations one can often notice that single fluorescent tubes have been removed or switched off by operators. Upon enquiry they cannot give plausible reasons for this but claim that a lower illumination level suits them better.

Some research has been done in the field of preferred lighting conditions as VDT workstations. Shahnavaz (180) carried out a field study in a Swedish telephone information centre. The operators could adjust the level of illumination on the work desk. The preferred mean illumination levels on the telephone directory were 322 lx during the day and 241 lx for night shifts with similar levels on the desk and on the keyboard.

Benz et al. (15) observed that 40% of VDT operators preferred levels between 200 and 400 lx and 45% levels between 400 and 600 lx.

van der Heiden et al. (84) observed at CAD workstations that many operators had switched off some of the lights, or drawn the curtains in front of the windows or lowered the blinds. The measured illumination levels could therefore be considered to be a preferred condition or a compromise between reduction of reflections and need of light. The results of 38 CAD workstations are given in Table 5.

*Recommended illumination levels at VDT workstations*

Snyder (190) says that ambient office illumination should be reduced to a level compatible with that of the VDT. Such levels are of the order of 200 lx and generally cause the office to appear dimly lit. This illumination is inadequate for reading hard-copy documents. To meet this insufficiency Snyder (190) recommends local lighting fixtures designed and arranged so as to avoid reflected or direct glare for other workers. The German DIN Standard (44) recommends an illumination level of 500 lx for offices with VDTs.

It is, however, not suitable to recommend merely one figure since the working conditions might differ from one job to another. For instance, Läubli et al. (120) observed in a field study that the preferences of operators working on data entry

**Table 6** Recommended illumination levels at VDT workstations.
*The lux figures refer to measures taken on a horizontal plane.*

| Working conditions | Illumination levels (lx) |
| --- | --- |
| Conversational tasks with well printed source documents | 300 |
| Conversational tasks with reduced readability of source documents | 400–500 |
| Data entry tasks | 500–700 |

tasks tended towards higher illumination levels than those engaged in conversational tasks.

General experience as well as several field studies lead to the recommendations given in Table 6.

## 5.3. Spatial balance of surface luminances

The distribution of luminances of large surfaces in the visual environment is of crucial importance for both visual comfort and visibility. In general, the higher the ratio of change or difference in luminance levels, the greater the loss in visibility, as was recently demonstrated by Boynton *et al.* (19) and Rinalducci and Beare (171).

*How to express luminance contrast*

Although there are many ways to define relative luminances, the most common one is simply to specify the ratio of two luminances. Many authors, though, prefer the definition adopted by the International Lighting Commission (CIE) according to the following formula:

$$C = \frac{L_O - L_B}{L_B}$$

where $C$ = contrast; $L_O$ = luminance of the target; and $L_B$ = luminance of the background.

*Sharp contrasts, e.g., relative or transient glare*

It is generally agreed that sharp luminance contrasts between large surfaces located in the visual environment reduce visual comfort and visibility, but the degree of acceptable contrast ratios is contested. Tolerable luminance ratios depend upon specific circumstances. Many factors are involved, such as size of the source of glare, its distance from the viewer's line of sight and the intensity of the general illumination in the room. Furthermore, results of experiments will also depend on

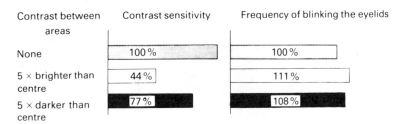

**Figure 17** Physiological effects of large surfaces with marked luminance contrasts located in the middle of the visual field.
*The contrast between an area covering 15% of the centre of the visual field and an area immediately adjacent to it is measured. According to Guth (76).*

whether quantitative visual performance or subjective visual discomfort is measured.

*The early study by Guth*

Among the earlier studies, the results of Guth (76) shall be mentioned here. He observed a decrease of contrast sensitivity and an increase of the eye blinking rate when the centre area of the visual field was 5 times brighter or darker than the adjacent area. Figure 17 shows the results of this study which is to some extent typical for what happens at VDT workstations.

According to these experiments relative contrast ratios of 1:5 in the middle of the visual field significantly impair the efficiency of the eye as well as visual comfort. If the adjacent areas are brighter than the centre area the disturbances seem to be more strongly felt than vice versa.

*General rules*

The following general rules are widely accepted:

1. All large objects and major surfaces in the visual environment should, if possible, be equally bright.
2. Surfaces in the middle of the visual field should not have a brightness contrast of more than 3:1 (see Figure 18).
3. Contrasts between the central and the marginal areas of the visual field should not exceed 10:1 (see Figure 18).
4. The working area should be brighter in the middle and darker in the surrounding field.
5. Excessive contrasts are more troublesome at the sides than at the top of the visual field.
6. Light sources should not contrast with their background by more than 20:1.
7. The maximum brightness contrast within the entire room should not exceed 40:1.

These rules also apply to the whole visual environment regularly covered by the operator's glance.

In everyday practice these guidelines are often neglected. As always one comes across contrasting elements which should be avoided in the visual environment, such as:

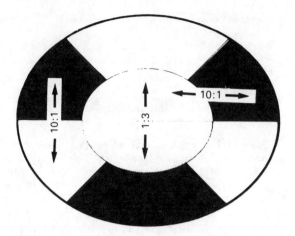

**Figure 18 Acceptable contrast ratios of brightness between different areas of the visual field.**
*Within the middle field, 3:1; within the outer field, 10:1; and between middle and outer field, 10:1.*

Bright walls
Dazzling white walls contrasting with dark floorings, dark furniture or black office machines
Reflecting tabletops

*Reflectances and colours*

The choice of colour and material is of major importance for a balanced distribution of luminance. The reflectance values generally proposed for offices without VDTs are as follows:

| | |
|---|---|
| Ceiling | 80–90% |
| Walls, blinds or curtains | 50–60% |
| Furniture | 25–40% |
| Flooring | 20–40% |

The special requirements for offices with VDT workstations will be discussed later.

If the above-mentioned reflectance figures are respected the actual choice of colour to be used in the office is mainly an aesthetic question. It must be added, however, that from the physiological point of view large surfaces painted in saturated colours cause distraction and eye strain, whereas colours of softer shades have proved more suitable. Some psychological effects of colours should also be taken into consideration and summarized here:

Blue and green seem further away, rooms appear larger, slightly colder and more restful.
Red and orange seem close, are warm and rather exciting.
Brown seems very close, rooms appear small, and many people find brown colours to be cosy.

# Ergonomic principles of lighting in offices

Table 7  Surface luminances and contrast ratios at 109 VDT workstations with conversational tasks, measured in different banks.

|  | Luminance ($cd/m^2$) | |
| --- | --- | --- |
|  | Median | 90% range |
| *Surface* | | |
| Source documents | 108 | 34–208 |
| Background of screens | 4 | 1–11 |
| Reflections on the screen | 17 | 5–120 |
| *Contrast ratio* | | |
| Source document to screen | 21:1 | 10:1–81:1 |
| Window to screen | 300:1 | 87:1–1450:1 |

*Surface luminances at VDT workstations*

The surfaces in the visual field and in the visual surroundings of a VDT operator are the screen, frame and enclosure of the display, the desk, keyboard and source documents, and other elements of the immediate environment such as walls, windows, ceiling and furniture.

The contrasts of surface luminances of VDT workstations with bright characters and dark background are often extremely sharp, similar to the contrast between the dark screen background and the bright source documents. Läubli et al. (120) measured the luminances at 109 VDT workstations with conversational jobs. The results are given in Table 7.

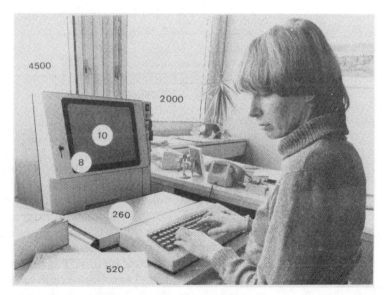

Figure 19  **Excessive luminance contrasts in the visual environment of a VDT operator.**
*The figures in the circles indicate the measured luminances expressed in $cd/m^2$. Screen and source document have a contrast ratio of 1:50, screen and window of 1:450.*

Nearly all the workstations showed contrast ratios exceeding the general rules for offices mentioned above.

Figure 19 illustrates an example of a very badly arranged workstation; unfortunately this picture cannot be considered an exception!

Bräuninger et al. (21, 22) measured the contrast ratios of surface luminances between screen background and source documents under the following laboratory conditions: an illumination level on a horizontal plane of 400 lx, and a luminance of the source document reaching 50 cd/m$^2$. Forty-seven VDT models[†] from 23 different European and American manufacturers were studied. The measured contrast ratios between source document and screen are reported in Table 8.

The results show that under favourable lighting conditions most of the VDTs will allow contrast ratios between screen and source document of less than 1:10, but only very few models gave contrast ratios of 1:3. It must be pointed out, however, that a contrast ratio of 1:3 or less would require a screen luminance of 16 or more cd/m$^2$. Such high luminance would then be definitely too high and generate poor sharpness and flickering characters.

Table 8  Contrast ratios between luminances of screen and source documents for 47 different VDT models.
*The luminance of the source document was kept constant at 50 cd/m$^2$.*

| Number of models | Contrast ratios |
|---|---|
| 31 | $<1:10$ |
| 5  | $1:10-1:15$ |
| 9  | $1:16-1:25$ |

*3:1 or 10:1?*

According to the general guidelines the contrast ratio between the dark background of a screen and a well illuminated source document should not exceed 1:3. It is obvious that this recommendation cannot be realized at the majority of VDT workstations with bright characters and a dark background. This situation induced several authors to reexamine the validity of the '1:3 rule'. Thus, Haubner and Kokoschka (81) studied the effects of different luminance ratios at the workplace on performance and subjective behaviour of users during an ordinary working day. The luminances on the paper document were 65, 325 and 1300 cd/m$^2$, on the screen 13 cd/m$^2$.

---

[†] Bräuninger et al. originally presented the results of 33 VDT models (22). Since then 14 others models have been evaluated to give the 47 VDTs taken into consideration here.

This gave luminance ratios of 5:1, 25:1 and 100:1. The time needed for writing down four-figure random numbers was used as the performance parameter. The authors concluded that no significant effects on the task performance were apparent up to luminance modifications of about 20:1 and that the required limit values of 3:1 and 10:1 are unnecessarily restrictive. This study can certainly not be considered to be the final proof for the acceptability of contrast ratios of 10:1 or more in the visual field. Other parameters, especially visual comfort and preference ratings, should also be taken into account. Rupp (173) critically examined the rules recommending contrast ratios of 3:1 between source documents and screen. He pointed out that the visual system's level of adaptation is determined by the luminance of the bright characters and not by an integrated luminance level or the background luminance level. If this assumption is correct one might be more concerned with matching the luminance of the bright characters with the source document background. This argument of Rupp's needs to be proved. The experts of the US National Academy of Sciences (153) claim that the evidence presented in support of Rupp's hypothesis is either unconvincing or inappropriate and conclude that the matter needs to be examined further.

*Recommended contrasts at VDT workstations...*

*The luminance contrast between dark screen (with bright characters) and source document should not exceed the ratio of 1:10. All other surfaces in the visual environment should have luminances lying between those of the screen and the source document.*

These recommendations are illustrated in Figure 20.

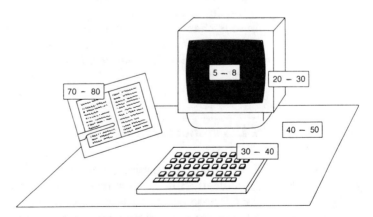

**Figure 20 Recommended reflectances for a VDT workstation with a dark screen and a bright workstation environment.**
*The reflectance is the percentage of reflected light related to the luminous flux falling onto the surface concerned.*

*... and offices with VDTs*

The reflectances for the entire office environment with VDTs should be slightly below those of a traditional office, mentioned earlier. A main reason for low reflectances is the risk of reflections from bright ceilings, walls and windows. The following reflectances can be proposed:

| | |
|---|---|
| Ceilings | 70% |
| Windows, blinds or curtains | 50% |
| Walls behind the screen | 40–50% |
| Walls opposite the screen | 30–40% |
| Flooring | 20–40% |

*Imagination of designers*

Some designers try very hard to come up with imaginative ideas in an attempt to design attractive furniture for offices. They visualize pitch-black office machines on a bright table or dark furniture neighbouring bright walls. Such designers don't care about ergonomic principles or balanced surface luminances.

The instructions for the designers of VDT workstations can therefore be summed up as follows:

*Select colours of similar brightness for the different surfaces, reject eye-catching effects with black and white contrasts, avoid reflecting materials and give preference to dim colours.*

*Contrast ratios for VDTs with reversed presentation*

In Chapter 3 a brief description of VDTs with reversed presentation was given. They are characterized by a bright screen background and dark symbols and such screens are to some extent comparable to printed source documents.

As yet there are only a few VDT models on the market with reversed presentation. The pros and cons of these terminals will be discussed later in Section 6.10. In this context it must be pointed out that VDTs with reversed presentation have a screen luminance ranging between 50 and 100 cd/m$^2$. It is obvious that such bright screens do not raise the problem of excessive contrast ratios with source documents or other bright surfaces in the visual environment, an undoubted merit!

## 5.4. Temporal uniformity of lighting

*Alternating bright and dark areas*

Even more disturbing than static luminance contrast are periodically fluctuating bright and dark areas in the visual field. These occur if the work requires the operator to glance alternately at a bright and a dark surface; if bright and dark objects pass by on a conveyor belt; or if moving parts of a machine are bright and reflecting.

As already mentioned the pupil and the retina of the eye can

# Ergonomic principles of lighting in offices

cope with changes of brightness only after a certain delay, so that fluctuating brightness leaves the eye either under- or over-exposed for much of the time. Hence such lighting conditions are particularly disturbing. In order to avoid fluctuating levels of brightness as far as possible:

*Cover moving machinery with an appropriate housing. Equalize brightness and colours along the main axis of sight.*

*Alternating viewing at VDTs*

In conversational VDT jobs the eyes scan from the dark screen to the bright source document and back again about 5—7 times per minute. The speed of alternation of the gaze does certainly not allow a proper adaptation of the eye. The necessary reduction of luminance contrasts between screen and source document has already been discussed above.

Another source of rhythmically oscillating luminances are fluorescent tubes, as discussed at the beginning of Chapter 5, and finally, the bright characters on VDT screens also produce an oscillating luminance — this phenomenon will be treated in the following chapter on the photometric characteristics of VDTs.

## 5.5. Appropriate lighting

*Lights as sources of glare*

Inadequate lights or lighting arrangements can be sources of glare which make viewing difficult and uncomfortable. *To avoid glare inside a room is one of the most important ergonomic considerations when designing offices.*

Figure 21 sets out the results of a classic piece of research by Luckiesh and Moss (134). Their test subjects carried out a visual task in which a light source of 100 watts was step by

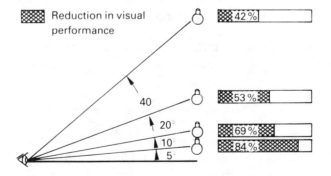

**Figure 21  Effect of glare on visual performance.**
*The grey blocks indicate the reduction in visual performance as a percentage of the normal performance (=without glare). Visual performance becomes worse the closer the light is moved to the optical axis. According to Luckiesh and Moss (134).*

**Figure 22 Unsuitable lighting in a drawing office.**
*The opalescent globes are sources of much glare. The dark floor contrasts too strongly with the white working surfaces and produces bright reflections of the lamps.*

step moved closer to the optical axis. Visual performance was gradually impaired.

As the figure shows, the lighting in a workroom must be carefully arranged. The lights in offices often generate both absolute and relative glare in addition to dazzling reflections on polished surfaces or on the glass surface of VDT screens.

*A bad example*

Figure 22 shows very unsatisfactory lighting conditions in a drawing office. In this office opalescent globes are being used which often come into the visual field of the draughtsmen as well as being reflected from the polished floor covering. This results in very sharp luminance contrasts, far exceeding the recommended maximum ratio of 10:1. The responsible architect must have been full of good intentions creating such a festive atmosphere...!

*Rules for lighting offices without VDTs*

On the basis of many studies as well as from experience the following rules can be deduced for all types of offices without VDTs:

1. *No light source should appear within the visual field of an office employee during his/her working activities.*
2. *All lights should be provided with shades or glare shields to prevent the luminance of the light source exceeding 200 cd/m$^2$.*
3. *The line from eye to light source must have an angle of more than 30° to the horizontal plane (see Figure 23). If a smaller angle cannot be avoided, e.g., in large offices, then the lamps must be shaded more effectively.*

*Ergonomic principles of lighting in offices* 49

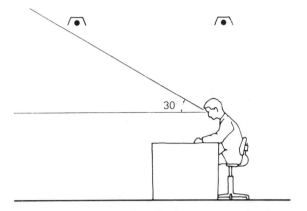

**Figure 23** The angle between the horizontal and the line from eye to lamp overhead should be more than 30°.

4. *Fluorescent tubes should be aligned at right angles to the line of sight.*
5. *It is better to use more lamps, each of lower power, than a few high-powered lamps.*
6. *To avoid annoying reflections from the desk surface the line from eye to desk should not coincide with the line of reflected light. This is illustrated in Figure 24.*
7. *The use of reflecting colours and materials on table-tops or office machines should be avoided.*

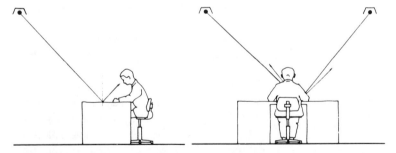

**Figure 24** Left: poor placing of a single lamp, so that its reflection falls into the employee's line of sight with the risk of glare. Right: the reflections of two lamps placed at either side are not in the line of sight, so reflected glare is avoided.

*Lighting in VDT offices*

Two ergonomic problems require special attention when lighting is to be designed for offices equipped with VDTs: sharp luminance contrasts between a screen and its surroundings must be avoided, and annoying reflections on the glass surface of screens must be reduced or eliminated. Having dealt with the problem of luminance contrasts earlier the thorny problem of reflections on the screen surface shall be treated here.

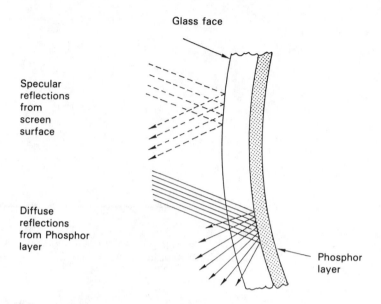

Figure 25   Specular and diffuse reflections from a video screen.

*Reflections on screen surfaces...*

The surface of a VDT screen is made of glass, and reflects about 4% of the incident light; this suffices to reflect clear images of the office surroundings such as lights, the keyboard or the image of the operator. Figure 25 shows how a video screen reflects light.

One part is reflected from the glass surface of the screen which produces a mirror-like reflection of the surroundings. The other part is reflected from the phosphor layer, producing a veiled and diffuse reflection of a light source.

*...produce glare or annoyance*

Bright reflections can be a source of glare and image reflections are annoying, to say the least, especially since they also interfere with focusing mechanisms; the eye is forced to focus alternately on the text and the reflected image. Thus reflections are also a source of distraction. Stammerjohn et al. (191) and Elias and Cail (51) observed that bright reflections on the screen are often the principal complaint of operators; the reflected luminances reached figures of between 3 and 50 $cd/m^2$.

In a recent experimental study Dugas Garcia and Wierwille (49) investigated the effect of reflected glare of screens on the reading performance of 10 subjects. Reading performance was assessed using reading time and correctness of answers as measures. Glare was produced by fluorescent light fixtures behind the subjects. With the help of a movable shield 'glare' or 'no glare' conditions could be provided. Glare was found to increase the amount of time required to read relatively easy

# Ergonomic principles of lighting in offices

passages and to decrease the amount of time required to read relatively difficult passages. The subjects judged the 'no glare' condition to be barely noticeable and the 'glare' condition to be quite disagreeable. In fact, it is well known that marked discomfort leading to strong complaint can be felt from glare long before any measurable change in performance of the task can be detected.

Figures 26 and 27 show common examples of reflections on screens at VDT workstations.

Figure 26 The reflected image of a window behind the back of the operator is superimposed on the screen text and disturbs reading.

Figure 27 A light source, reflected on the screen, generates glare and impairs reading.

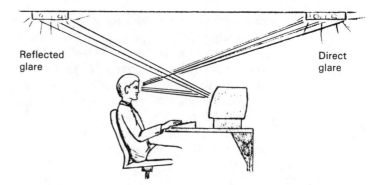

Figure 28 Light sources behind the operator represent a risk of reflected glare; lights in front of the operator cause direct glare.

*Positioning of VDT workstations*

*The most effective preventive measures are the adequate positioning of the screen with respect to lights, windows and other bright surfaces.* (Other protective measures, such as adjustment of screen angle and anti-reflective devices on the screen surface will be discussed later in Section 6.8.)

If the light source is behind the VDT operator it can easily be reflected onto the screen and cause reflected glare. If it is in front of him it can cause direct glare. These conditions are illustrated in Figure 28.

*...with respect to light fixtures*

Light fixtures directly above the operator can dim the characters with blurred reflections generated in the phosphor layer. Thus it is preferable to install the light fixtures parallel to and on either side of the operator – screen axis.

*...with respect to windows*

In offices, windows play a role similar to lights: a window in front of an operator disturbs through direct glare, behind it

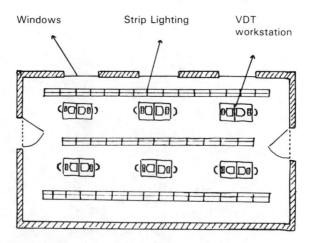

Figure 29 VDT workstations should be arranged at right angles to the window.
*Plan of an office layout with one window wall.*

produces reflected glare. *For this reason the VDT workstation must be placed at right angles to the window.* In offices with only one or perhaps two parallel window fronts this is an efficient protective measure. Figure 29 shows an ideal arrangement of VDT workstations in relation to a single window wall.

*Cover windows*

In offices with two or more window walls some form of window covering must be used. Windows must also be covered at night because the reflections of interior office lights may cause glare. Two types of window coverings can be useful:

*Louvers or mini-blinds.* Horizontal as well as vertical louvers can be used. Their purpose is to occlude the window on a bright day or to absorb light from indoor sources at night.
*Curtains.* They are efficient; preference should be given to material of a low reflectance of about 50%.

Finally, there is the possibility of placing intermediate screens between the VDT workstation and bright windows. Such a screen should not have a reflectance higher than about 50%.

*Appropriate light fixtures*

The best light fixtures for offices with VDTs are not the same as those for traditional office environments. Fixtures which provide a great deal of mainly horizontally directed light should be avoided since such light illuminates the vertical screen and generates reflections on it. It is advisable to use fixtures which provide a confined primarily downward distribution of light, with either built-in louvers, curved mirrors or prismatic pattern shields. The luminous flux angle should not exceed 45° to the vertical. Suitable and well

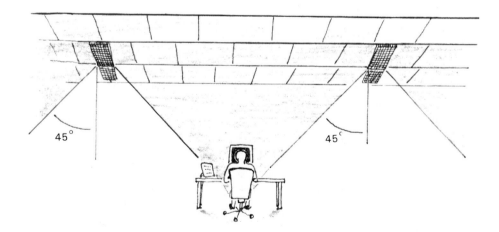

**Figure 30** Ceiling lighting with a prismatic pattern shield generating a cone of light with an angle of 45° to the vertical.

arranged light fixtures are shown in Figure 30. Such fixtures cause neither direct nor reflected glare since both screen and keyboard are in a shadow area.

*Indirect lighting*   Some lighting engineers suggest suspending the lighting from the ceiling, thus permitting the lowered fixtures to direct the greater part of the light upwards to the ceiling. Standard lamps emitting all light upwards towards the ceiling and the upper part of the walls are also used in some offices. These lighting systems may produce a pleasant aesthetic effect but have the drawback of producing bright ceilings and walls, which, in turn, may cause bright reflections on the screen.

*Task lighting*   Given a low general illumination level supplementary lighting may be provided for the source documents, especially when they have a low readability. It is important that such task lighting be confined to the area of the source documents. In order to avoid reflections on neighbouring VDT workstations the task lamps should not be transparent at the sides.

# 6. Visual Strain and Photometric Characteristics of VDTs

## 6.1. Eye complaints of VDT operators

*Field studies*

As mentioned in Chapter 1, the expansion in the use of VDTs has been accompanied by complaints by many VDT operators about visual strain and physical discomfort. These reports, initially not very credible, were followed by more systematic field studies in the late 1970s to check the authenticity of the complaints. Such systematic research has been carried out in England (88), France (51), FR Germany (27), Japan (160), Italy (61), the Netherlands (74), New Zealand (35), Sweden (75, 99, 104, 164), Switzerland (84, 120, 141) and the USA (5, 41, 177, 185, 186, 192, 193).

The majority of these field studies have been surveyed by Dainoff (42), Läubli and Grandjean (123) and Helander *et al.* (86).

*Controversial results*

Seventeen of these field studies disclosed an increased incidence of visual discomfort among VDT operators, with symptoms such as visual fatigue, pain in the eyes, burning or itching eyes, blurred or double vision and other troubles. Four of the studies, however, did not confirm these results; the frequency of complaints among VDT operators did not significantly exceed that of control groups.

All studies used either a standardized interview procedure or some kind of self-rating questionnaire. However, the results are in most cases not comparable, since the survey methods as well as the studied professional groups differ greatly from each other.

*Objections*

A few surveys failed to include control groups; in most cases non-VDT groups were used as controls but these again differed from the VDT groups not only in the use of VDTs but also in many other respects. The expert panel of the US Academy of Sciences (153) analysed six of the studies very carefully (35, 41, 51, 99, 185, 186), and objected to the incomplete statistical analyses and the missing connections with physical or mental loads, and criticized the use of inappropriate control groups. The final conclusion of the panel was formulated rather severely: "Thus it is not possible to determine from existing studies to what extent complaints reported by VDT operators have resulted from the VDT itself as opposed to such factors as workstation or job design. Our review of field surveys puts down that existing studies have not been able to establish whether VDT work *per se* produces more visual complaints than comparable non-VDT work". Helander et al. (86), who analysed 10 studies related to visual discomfort, arrived at a similar conclusion; they criticize the lack of scientific rigour which reduces the value of many of these studies.

*The magic of control groups*

The problem of control groups is indeed very intricate. The introduction of VDTs is normally accompanied by changes in task design, organization, speed of work, environment and especially major differences in performance and productivity. For these reasons it will hardly be possible to get a VDT group and a non-VDT group which will differ only in the use (or non-use) of VDTs. Table 9 shows the results of two surveys, carried out by Läubli et al. (120) and van der Heiden et al. (84, 85). In these surveys the same questionnaire was used in which the subjects could rate the symptoms with "never", "seldom", "occasionally" or "almost daily". In Table 9 only the item "tired and strained eyes" is taken into consideration.

Table 9 Incidence of complaints of eye strain rated as "almost daily".

| Groups | Reference | n | VDT use | Eye strain (%) | p |
|---|---|---|---|---|---|
| Data entry | (120) | 53 | + | 19 | |
| Payment transactions | (120) | 109 | + | 27 | 0.01 |
| Payment transactions | (120) | 55 | − | 6 | |
| CAD work | (85) | 69 | + | 15 | 0.05 |
| Technical drawings | (85) | 52 | − | 4 | |
| Full time typists | (120) | 78 | − | 18 | |

$n$ = Number of subjects per group; + = with use of VDTs; − = job without VDT.

*Incidence of strained eyes*

The following conclusions can be drawn from these results:

1. The groups dealing with payment transactions in banks and the groups performing CAD work showed a significantly higher incidence of strained eyes than the corresponding control groups. The compared groups were similar with respect to age and sex.
2. The occurrence of strained eyes among data entry groups

using VDTs is of the same order of magnitude as among full-time typists. The difference to the group of payment transactions with VDT is not significant. The occurrence of strained eyes is of similar frequency in all three groups.

*The problem of comparing jobs*

The payment transaction groups and the CAD groups — both using VDTs — differ from their control groups in many respects. VDT operators engaged in payment transactions carry out nearly 10 times as many orders per day as the control group doing the same job without VDTs but both groups perform a great variety of activities, such as getting up to check documents, searching in lists, making telephone calls, reading and typing. CAD operators carry out technical drawing with the computer, watching the progress on the screen for the greater part of the time. The control group does the same job on traditional drawing tables. Here again the job is the same but the work load, especially with regard to speed and performance, is certainly greater for the CAD group than for the control group.

*Strenuous full-time typing and data entry*

Data entry using VDTs and full-time typing are, on the other hand, jobs which are defined by a high rate of daily key-strokes. For the former Läubli (123) estimated 60 000, for the latter 50 000 strokes per day. The tasks require a constantly high degree of attention, and interruptions are rare. Hence it is not surprising that both groups show a frequent occurrence of complaints. Läubli (123) assumes that increased general fatigue can be accompanied by 'tired and heavy eyes'. The results of these studies illustrate the problematical aspects of comparable control groups.

*The studies by Starr*

In this context the studies by Starr et al. (192), Starr (193) and de Groot and Kamphuis (74) will be discussed briefly, none of them confirming the above results.

In the first survey (192) a self-administered questionnaire was completed by 145 telephone operators who spent their entire working time answering customer requests for telephone numbers using VDTs for the research and by 105 control subjects who performed the same task but retrieved the requested entries from printed paper records. The two groups did not markedly differ in their experience of work-related physical discomfort and job satisfaction. In the second survey (193) Starr used the same questionnaire to study 211 service employees of the telephone company handling requests from subscribers (new telephones, changes in telephone service, billing enquiries, etc.) with the aid of VDTs; the corresponding control group of 148 subjects did the same services without a VDT. This study, too, revealed few differences in the probability and intensity of visual discomfort. Differences were found with regard to job satisfaction

**Table 10 Visual discomfort of telephone operators on working days of previous month.**
*According to the studies by Starr et al. (192) and Starr (193).*

| Visual discomfort | Operators reporting discomfort (%) | | | | | |
|---|---|---|---|---|---|---|
| | VDT A | Paper A | p | VDT B | Paper B | p |
| Blurred vision or difficulty focusing | 52 | 42 | n.s. | 46 | 32 | 0.01 |
| Double vision | 10 | 8 | n.s. | 5 | 1 | n.s. |
| Burning, tearing or itching eyes | 61 | 55 | n.s. | 56 | 53 | n.s. |
| Sore eyes | 65 | 54 | n.s. | 35 | 41 | n.s. |

VDT A and Paper A: directory enquiry operators and their control group. VDT B and Paper B: telephone operators handling requests from residential customers and their control groups. $p$ = Significance; n.s. = not significant.

and concerns about job security, clearly favouring VDT users. The occurrence of visual discomfort in both studies is reported in Table 10.

*Visual discomfort of telephone operators*

The results point to a slight tendency towards increased visual discomfort among the VDT groups but there is only one significant difference. Understandably, Starr deduces from these studies that a mere substitution of VDTs for paper documents is not associated with a syndrome of adverse physical and psychological effects. He believes that a VDT does not stress the visual system more than analogous near-vision work done without a VDT. In fact, the traditional work of telephone employees is strenuous, given the high working speed requiring a constantly high degree of attention. Starr (193) points out that the documents are often poor quality computer print-outs, and may even be handwritten, and in pencil rather than ink. Hence it is not surprising that visual discomfort was very frequent in both control groups. The Starr studies on telephone operators again raise the problem of reliable and representative control groups. Starr (193) himself is very cautious when he states: "It is clearly unreasonable to suggest that the equivalences between VDTs and paper will hold to the same extent in all jobs and under all conditions. Not all environmental conditions are reasonable, however. It shouldn't require much imagination to contrive, nor effort to find, conditions that are highly favorable to paper and unfavorable to VDTs".

The third study, which confirms Starr's results, was also carried out on telephone enquiry personnel: de Groot and Kamphuis (74) compared the results of a questionnaire and some physiological measurements on 43 subjects just before, just after and two years after the introduction of VDTs. The survey 'before' was therefore the control condition of the investigation. Between 30 and 40% of the subjects reported that they suffered from eye disturbances. But number, type and severity of complaints did not change over the years. The measurements of visual acuity, of accommodation and flicker fusion frequency showed no deterioration other than ageing effects. The general satisfaction with the new, automated system was high and most of the subjects worked only part-time. The authors believe that these two factors could have mitigated the complaints. These results, similar to those of Starr (192, 193), suggest that the traditional work in telephone enquiries remains strenuous despite the introduction of VDTs. This explains why visual complaints did not become fewer.

Howarth and Istance (88) recently confirmed to some extent the above results. They conducted a field study on four groups of workers: two VDT groups (data preparation and word processing) and two non-VDT groups (typing and clerical work). The groups were studied for a period of five days. Subjective ratings of symptoms were taken at the beginning and at the end of each day. The VDT groups made more complaints both in the morning and at the end of the day. When the subjects were questioned on "change over the day" no significant differences attributable to VDT use were found.

*Summary of surveys on visual discomfort complaints*

Summarizing all the cited field studies the following statements can be made:

The majority of the surveys (17 studies) showed an increased incidence of visual discomfort. The fact that many of these studies were lacking in scientific accuracy reduces the value of the findings but certainly no proof is furnished that visual discomfort is non-existent among VDT operators. Three studies on telephone operators with VDTs did not reveal higher incidences of visual discomfort than control groups doing the same work without VDTs. The same result was obtained comparing VDT operators with full-time typists (88) (120).

The controversial results might to some extent be explained by the choice of control groups. If the control groups are engaged in traditional office work with low productivity and a great variety of activities they seem to be much less affected by visual discomfort than the VDT groups. If, on the other hand, the control groups do strenuous work at high speed and under poor reading conditions the complaints of visual discomfort might be as frequent as among VDT groups. These considerations demonstrate that control groups in field studies are often very problematical. To put it cynically, with field studies every hypothesis can be proved, it all depends on the choice of an appropriate control group!

*The symptoms of visual discomfort*

What kinds of visual discomfort were reported and what is their implication?

**Table 11 Factor analysis of symptoms of visual discomfort reported by a questionnaire.**
*From a survey by Läubli et al. (120).*

| Factor | Symptom | Loading |
| --- | --- | --- |
| Factor 1 | Pains | 0.71 |
| 81% of extracted variance | Burning | 0.66 |
| | Fatigue | 0.64 |
| | Shooting pain | 0.53 |
| | Red eyes | 0.49 |
| | Headaches | 0.42 |
| Factor 2 | Blurring of near sight | 0.79 |
| 19% of extracted variance | Flicker vision | 0.62 |
| | Blurring of far sight | 0.45 |
| | Double images | 0.45 |

Läubli et al. (120) carried out a factor analysis with a varimax rotation of all variables asked for in the questionnaire about eye impairments. Two factors were extracted, with a clear distribution of the variables between them. The results are reported in Table 11.

Läubli assumed that factor 1 could be related to eye fatigue or eye irritation and factor 2 to accommodation strain. The listed symptoms were identical or at least very similar in all the field studies mentioned above. *There is a general agreement that the reported symptoms of visual discomfort are reversible and functional troubles and not persistent eye injuries.* Nevertheless, visual discomfort may persist for several hours. Läubli et al. (120) remarked that some VDT operators as well as full-time typists reported visual discomfort when watching television or when reading during leisure time.

*Visual tests*

Several authors tried to get more objective results by checking the effects of VDT work on visual functions in an attempt to find physiological correlates of visual fatigue. The general procedure was to take initial measurements of visual tests, require the subjects to carry out a task with the VDT or with hard copy material as reference and then to repeat the visual tests. In field studies the VDT groups were compared with control groups doing similar work without VDTs.

Accommodation, convergence, phoria, visual acuity, critical fusion frequency (CFF) and contrast sensitivity have been tested by several authors. In several cases the validity of the reported results was to some extent limited. Some of the publications were analysed by Helander et al. (86), who stated that they are by no means conclusive.

*Accommodation*

The visual function studied in greatest detail was accommodation capacity or resting accommodation in the dark. These studies were based on the assumption that the ciliary muscle of the eye fatigues and should consequently be less able to adjust flexibly to the requirements of the observer. Some of these investigations are summarized below.

Östberg et al. (165) used a field laser optometer to measure accommodation under both illuminated and dark conditions. Three groups of subjects were tested before and after work sessions with displays. Air traffic controllers showed significant changes in their resting point of accommodation (dark focus point), which moved closer to the head after two hours of work. The VDT groups disclosed no significant changes. Gunnarson and Soederberg (75) measured near points of accommodation and convergence. When VDT routine work was intensified an increase in the near point of accommodation and convergence was recorded. The near point increased by an average of about 3.5 cm during a day of intense VDT work, compared to 1.2 cm on a normal working day. Elias and Cail (51) recorded a significant increase of the near point in a data-acquisition task, but no significant change in a conversational VDT group.

*Accommodation changes not confirmed*

A number of recently published studies have not confirmed changes of accommodation capacities due to VDT work.

Hedman and Briem (82) measured accommodation and convergence near points, focusing accuracy and dark focus before and after work at a VDT and in two control conditions. The results showed no differential effects between the three tasks, either in near points of accommodation and convergence or in focusing accuracy. The authors observed that the most important variable in this type of study is the age of the operator. These results were recently confirmed by Nyman et al. (162) who could not find any difference in accommodation and convergence capacity between VDT operators and a reference group.

*Visual functions in laboratory experiments*

Interesting laboratory experiments were carried out by Gould and Grischkowsky (64): 24 subjects proofread from a VDT on one day and from hard copy on another day to investigate the role a VDT, apart from the work itself, plays in any changes in people's performance, feelings or vision. The VDT used in this experiment had good photometric qualities and the hard-copy material was also of good legibility. The task of the subjects consisted of identifying misspelled words in excerpts from magazines and newspapers. Contrast sensitivity, visual acuity, sensitivity to flicker and lateral phoria were the visual functions tested at the beginning of each day and at the end of each work period. Throughout the day no change in participants' proofreading performance (speed and accuracy), visual comfort, feelings or visual functions could be attributed to using VDTs. Any changes in participants' performance, feelings and visual functions that occurred during the day were due to the work itself and were unaffected by the subjects' use of VDT or hard-copy material. However, subjects did proofread hard copy about 20 to 30% faster than VDTs. A follow-up pilot work showed that the differences in the time needed to circle a misspelling on paper versus the time necessary to point at it on the screen could account for only 1/10 of the difference.

*Conclusions from Gould and Grischkowsky*

The authors point out that their study had some limitations: it lasted for only two days and the task was to search for misspellings; it was carried out under good ergonomic conditions (lighting, seating, lack of glare and good quality display). The authors were aware that in practice the organization of work is usually affected by the use of a computer. It is possible that people are placed in an unfavourable work environment (e.g., with poor VDTs) and asked to do jobs that can lead to adverse effects. The authors conclude as follows: "*Ergonomics, thoughtfully designed and evaluated user interfaces, and humane management practices are required not only to arrive at exemplary workplaces, but also to achieve personally fulfilling work design and organization*". This is certainly a very constructive conclusion coming from representatives of a large computer company.

Two laboratory studies carried out by Nishiyama *et al.* (161) and Gyr *et al.* (77) confirmed that a 3 hour reading task provoked the same changes of visual functions whether a VDT with a sufficient refresh rate or a normal printed text was used.

*Performance with VDTs*

Two similar experimental studies with the aim of checking VDT effects on reading performance were conducted by Mutter *et al.* (148) in Canada, and in a follow-up replication by Kruk and Mutter (118) 32 subjects read continuous text for 2 hours. Half of them read from a television screen (Videotex), half read from a book. The book copy was arranged as 40 rows per page and approximately 60 characters per row (400 words per page). In the Videotex copy there was a maximum of 18 rows of text with 39 characters per row and approximately 120 words per 'screen-page'. The displayed $5 \times 8$ matrix height measured 1 cm and the subjects were seated about 2.5 m from the screen. Subjects experienced very little nausea or headache in either condition. A small amount of dizziness, fatigue and eyestrain was produced by reading, but there were no significant differences between reading the book and reading the Videotex, nor was there a difference in comprehension scores. Videotex subjects read 28.5% more slowly than book subjects. Several possible reasons for this effect were explored in the follow-up study (118). Extra time to fill

*Reading is slower on Videotex*

the screen had no significant effect on reading time in the Videotex condition. Similarly, varying the contrast ratio of the video image and the distance between screen and reader had no effect on reading speed. The format of the Videotex had some slowing-down effect, but it could not account for the difference between book and video condition. However, an experiment with single and double spacing revealed a marked increase of reading speed with large interline spaces. In fact, the distances between lines in single spacing were much smaller than generally recommended. As mentioned in Section 4.7 ('Physiology of reading'), the leftward line saccades require sufficiently large spaces between the lines; interline spaces of 100 to 150% of the character height are generally recommended (see Section 6.9). The conclusion of the authors, that tight vertical spacing in the Videotex condition may contribute to slow reading, is certainly correct.

*Correlates of visual discomfort*

As mentioned earlier the field studies on visual discomfort yield controversial results and the question to what extent the complaints depend on the VDT *per se* or on other factors remains open. Those studies which reveal significant relationships between physical characteristics of VDTs on the one hand and symptoms of visual discomfort on the other are certainly more conclusive. These results shall be discussed in the following.

Already Dainoff *et al.* (41) have observed more complaints about glare and eyestrain (60%) in a group using an old VDT model than in another group working with an improved model (36%). The differences were suggestive, though they failed to reach statistical significance.

*Screen readability might be bothersome*

Stammerjohn *et al.* (191) conducted an ergonomic evaluation of VDT workstations in five different establishments. A majority of the operators surveyed found a number of factors to be bothersome, including screen readability, reflected glare, screen brightness and flicker. A significant relationship existed between visual function complaints and the employees' rating of glare and screen flicker. Smith *et al.* (184) conducted a survey in a large newspaper company to identify the relationship between ocular and somatic complaints and VDT use. Poor visual clarity and readability of the VDT screen explained the plurality of work-associated symptoms.

*Luminance contrasts and oscillation...*

The most striking connections were found by Läubli *et al.* (120, 122) in the survey already described (see Tables 9 and 11). The oscillating luminance of the screens as well as the luminance contrast ratios between screen and environment were determined. High luminance contrasts were associated with an increased incidence of visual discomfort. Oscillating luminance was defined as the base of the oscillation divided by

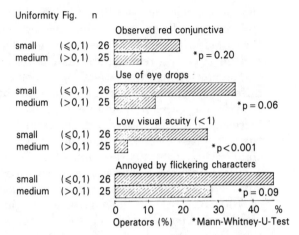

**Figure 31** Incidence of objective and subjective symptoms of visual discomfort related to the extent of luminance oscillation of screen characters.
*51 operators engaged in a conversational task. n = number of subjects per group with low or with high luminance oscillation. According to Läubli et al. (120).*

*...might cause visual discomfort*

the peak during one cyle. Increased oscillating luminance of screen characters was associated with lower visual acuity and with a higher incidence of subjective and objective symptoms of eye irritation including more frequent use of eye drops. Some of these results are shown in Figure 31.

*Objections related to screen quality*

A large number of operators were aware of the poor quality of the VDTs. These objections are reported in Table 12.

*Läubli's comparative study of two VDT makes*

The most striking result of Läubli's survey was the finding of a significant difference in visual discomfort between two operator groups using different VDT makes. The two groups were comparable with respect to age, sex and work (payment transactions). One group of 54 operators used VDT 'B' with good photometric qualities while the other group of 55 operators had VDT 'G' with poor photometric display

**Table 12 Objections of 109 VDT operators engaged in conversational tasks.**
*According to Läubli et al. (120).*

| Objection | Operators (%) |
|---|---|
| Disturbing reflections | 40 |
| Disturbing flicker | 26 |
| Blurred characters | 5 |
| Poor legibility of characters under normal lighting conditions | 9 |
| Poor legibility when looking from source document to screen | 15 |
| Disturbing contrasts in the visual field | 40 |
| Response time too long | 18 |

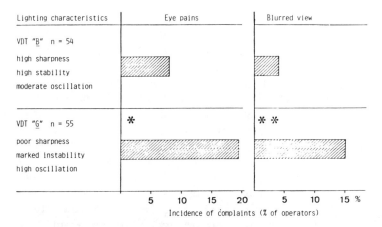

Figure 32 The incidence of complaints of "almost daily" eye troubles in two groups of operators using two different makes of VDTs.
n = *number of operators per group.* \* = $p < 0.05$. \*\* = $p < 0.01$. According to Läubli et al. (122).

qualities. The comparison of two symptoms of visual discomfort among the two groups is shown in Figure 32.

A correlation is not *eo ipso* a proof for a causal relationship. This must also be taken into account when discussing the findings of correlates between physical characteristics of VDTs and incidences of visual discomfort. Nevertheless, the described findings strongly suggest that an important part of visual discomfort among VDT operators might be caused by poor photometric qualities of the displays and by inadequate lighting conditions at the workstation.

## 6.2. Photometric characteristics of displays

*Differences between printed text and VDT*

The first question which arises when visual comfort at VDTs is at stake is: what may be the difference between reading a printed text and reading a text displayed on a VDT? Compared with a printed text, the main and often observed photometric differences are as follows:

> In most VDTs the characters are luminous on a dark background.
> The luminous characters do not exhibit constant luminance but a continuous repetition of light flashes. This means that the characters' luminance is of an oscillating nature.
> The text often displays low sharpness.
> The luminance contrast between characters and the screen background may be low.

The characters may exhibit movement and instability.
The text sometimes shows an insufficient geometric design of characters and typeface.

*Importance of photometric characteristics for visual comfort*

It must be assumed that sharpness, contrast and poor character design reduce the speed and precision of accommodation while contrasts of surface luminances as well as a flickering typeface may disturb the adaptation of the retina. *One important consequence of disturbed accommodation and adaptation might be lower legibility and occasional visual fatigue.* Good legibility is certainly a *conditio sine qua non* for visual comfort.

Some of the photometric characteristics, important for visual comfort, will be discussed in the following sections.

## 6.3. Equipment and methods to measure photometric qualities of VDTs

Fellman *et al.* (55) and Bräuninger *et al.* (20, 22) have developed equipment and a methodology to measure and evaluate the photometric characteristics of VDTs. Bräuninger (21) gave a full description of these studies in his thesis. Only the principles of the equipment shall be mentioned here. A microscope picks up the luminance of a small dot with a diameter of 0.1 mm inside a bar of a character, feeds it to a photomultiplier which amplifies the signals and transfers them to an oscilloscope, a DC voltmeter, an AC voltmeter, a Fourier analyser and a linearcorder. The band width of the system is 1 MHz. The luminance oscillation of the characters was measured on a display surface of $5 \times 7$ cm by a camera. Figure 33 shows the block diagram of the measuring arrangement and Figure 34 the microscope and the photomultiplier in front of a VDT screen.

*The procedure of Bräuninger*

The luminances of the various surfaces at the VDT workstations were measured with a Tektronix instrument. All measurements were carried out under standardized lighting conditions: indirect constant light with a vertical luminous flux of 400 lx and a horizontal luminous flux of 160 lx. If necessary the measurements were conducted with an adjusted luminance of the characters, the so-called 'preferred luminance'. These preferred figures were between 6 and 70 cd/m$^2$ and were assessed by the experimenters. In offices, operators adjusted luminances to between 9 and 77 cd/m$^2$, the mean value being 33 cd/m$^2$ (31).

*The Modulation Transfer Function*

Another procedure for quantifying the photometric qualities of the displayed image is through the Modulation Transfer Function (MTF). Snyder (190) applied the MTF to measure and evaluate the image quality of VDTs. This

Visual strain and photometric characteristics of VDTs 67

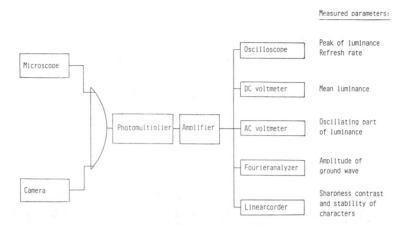

Figure 33 The block diagram for measuring various photometric parameters of VDTs.
*According to Bräuninger (21).*

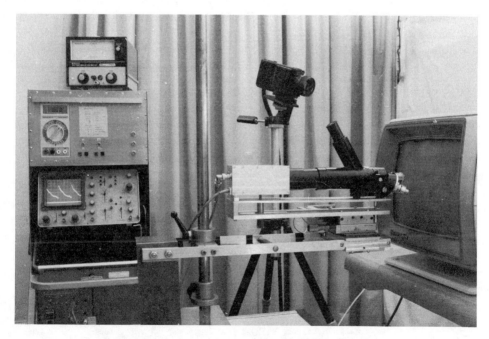

Figure 34 Microscope and photomultiplier in front of a VDT screen.
*According to Bräuninger (21).*

procedure is based on a Fourier transformation of the recorded luminance contrast between bright bars and dark screen background. He explains his application of MTF as follows:

*Ergonomics in Computerized Offices*

"This function is simply the contrast (modulation) expressed as a function of the size of the bars on a sine-wave grating, with increasing spatial frequency (e.g. cycles per unit visual angle) denoting decreasing bar width. More modulation per unit spatial frequency indicates greater contrast and perceived sharpness to the displayed image". More details are given in a Technical Report by Snyder (189). With the MTF, Snyder determined character contrasts, sharpness and the effects of glare on image quality (30, 71).

*The procedure of Menozzi and Krueger*

Recently a new approach has been adopted by Menozzi and Krueger (140), who used a special type of camera. As opposed to a microphotometer, this camera has a spatial resolution of about 1 minute of arc, which corresponds to the resolution of the eye. In the focal plane of the camera a line of about 1000 integrated photocells (charge coupled devices) is mounted horizontally, permitting the measurement of the luminance as a function of the horizontal path. A computer-controlled $x$-$y$ table allows the recording of two dimensional pictures of luminance distribution. Figure 35 shows the outlines of equal luminances of a displayed capital letter 'H'.

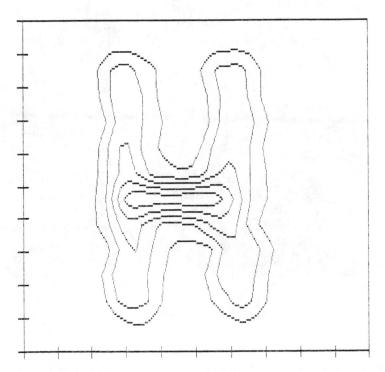

**Figure 35 Outlines of equal luminance of a displayed capital letter 'H' on a VDT.**
*On the abscissa and on the ordinate the divisions correspond to 0.73 mm (140).*

Furthermore, a photodiode — also mounted on the camera — is used to record time-dependent luminance.

With this device the authors determined the contrasts, the sharpness and the degree of oscillation of displayed characters.

## 6.4. Oscillating luminances of characters

As already mentioned in Chapter 3, the light of characters is composed of flashes with the frequency of the refresh rate of the CRT. Refresh rate and phosphor persistence determine whether the luminance oscillation is perceived as flicker or as constant light. In Section 5.1 it was stated that visible flicker of fluorescent tubes is extremely annoying and causes discomfort.

*Refresh rates and CFF*

The majority of VDT makes have refresh rates of 50 or 60 Hz. These rates seem to be in the critical range where some operators might already begin to perceive flicker. Läubli et al. (126) and Gyr et al. (77) used a simulated CRT, generating various oscillating frequencies between 0 and 180 Hz with chopper discs. The generated oscillating luminances were very similar to those of VDTs with a decay time of 5 ms to 10% of the peak luminance. The threshold of perceived flicker, called the critical fusion frequency (CFF), was measured on 29 subjects with this equipment. The results disclosed a range of individual CFFs of between 40 and 56 Hz. One restriction must be pointed out: in a CRT the electron beam moves down line by line, whereas in the simulated CRT used the total image of characters on the screen is turned on and off simultaneously.

Kelly (102) determined the CFF as a function of the fundamental and the mean luminance of the oscillations. On the basis of these studies one should expect CFF levels above the range of 50 to 80 Hz for VDTs.

*CFF with reversed displays*

Bauer et al. (8) and Bauer (10) determined for 31 subjects, aged between 18 and 48, the CFF on a VDT with reversed presentation (bright screen background). The phosphor $P_4$ had a decay time of 0.15 ms to reach 10% of the maximum luminance. The range of the measured CFF was between 55 and 87 Hz with a mean figure of 73 Hz.

Similar results were obtained by van der Zee and van der Meulen (213), who measured the CFF for 24 subjects watching a large area on a bright display. The 75th percentile of CFF was determined for large-area flicker to be between 75 and 85 Hz according to the viewing distance and the brightness of the display area. The authors concluded that a flicker-free display can be guaranteed for all practical cases of screen

*Angles of flicker perception*

luminance and viewing angle only if the field repetition frequency is increased to at least 92 Hz.

Isensee and Bennett (97) did not determine the CFF but the angle away from the CRT at which the subjects first noticed flicker. For this measurement the subjects swivelled the chair away from directly facing the CRT until reaching the point where flicker first became noticeable. The experiments were conducted on 21 subjects with a VDT with a 60 Hz refresh rate and a $P_4$ (white) phosphor. Furthermore, the subjects rated the level of flicker that was present on a scale from 0 (pleasant) to 6 (intolerable). The results showed an average angle of 25° (with 100 lx and 65 cd/m$^2$ luminance) at which the average flicker discomfort rating reached a figure of 3.2, which lies on the borderline between comfort and discomfort. Flicker was noticed at a smaller angle and was rated more uncomfortable with lower levels of ambient light and with higher screens luminance. Finally, flicker was perceived at a smaller angle for reversed than for normal screens.

The fact that flicker is perceived at rather large angles sideways of the VDT screen might to some extent explain why some data entry operators (who mostly watch the source documents) also complain about visual discomfort.

The few mentioned studies on flicker perception at VDTs do not yield consistent and conclusive results. As pointed out in Chapter 3, the CFF depends on two factors: the refresh rate of the CRT and the persistence of phosphor. Unfortunately the studies on flicker of VDTs did not consider the relationship between these two factors.

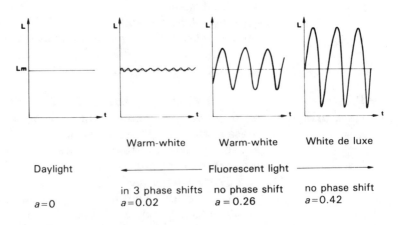

**Figure 36 Degrees of oscillation of natural light (far left) and of three types of fluorescent light.**
L = *luminance*; L$_m$ = *mean luminance*; t = *time*;
a = *degree of oscillation according to the formula mentioned below. According to Fellmann (55) and Bräuninger (20, 22).*

*Measurement of the degree of oscillation*

This is why Fellmann and Bräuninger (20, 22, 55) based their measurements of the degree of oscillation on the concept of Kelly (102), taking into account the amplitude as well as the mean luminance of oscillation. (This procedure indirectly includes refresh rate and persistence of phosphor.)

The degree of oscillation is, in fact, determined by recording the amplitude of the oscillation over the mean luminance. The measuring procedure was based on the following formula:

$$a = \frac{1}{L_m} \sqrt{\sum_{n=1}^{20} A_n^2 \, \text{eff}}$$

where $a$ = degree of oscillation; $L_m$ = mean luminance; and $A_n$ eff = amplitude of the fundamental wave and of first harmonics of a Fourier transformation.

*Degrees of oscillation of fluorescent lights...*

With this method Fellmann and Bräuninger (20, 22, 55)[†] measured the degree of oscillation $a$ for 45 different VDT models. In order to get a comparison they also measured $a$ for three types of fluorescent tubes. These results are shown in Figure 36.

They reveal marked differences in the amplitudes of oscillation as well as in the determined figures of $a$, depending on the type of fluorescent light. As mentioned in Section 5.1 the oscillation of fluorescent tubes is efficiently reduced by means of two or more phase-shifted tubes arranged in one lighting unit. This effect of phase-shifted equipment has already been well illustrated in Figure 15 and the results for the degree of oscillation $a$ are shown in Figure 36.

*...and of 45 VDTs*

The degree of oscillation $a$ was determined for 45 models from 23 different VDT manufacturers (13 US and 10 European makes). The measurements were made by taking the luminance of a single dot inside a bar of a character and of a larger surface of 5×7 cm including several lines and many characters. Each measurement was carried out with the preferred luminance (see Section 6.3).

The results obtained from the surfaces (5×7 cm) with an environmental illumination of 400 lx are reported in Table 13.

The measurements revealed great differences in the degree of oscillation. Since the refresh rates were of the same order of magnitude for all VDTs — namely 50 or 60 Hz — the differences are mainly due to the varying phosphor persistence. In fact, the fastest phosphor had a decay time of 2 ms whereas the slow ones showed decay times of more than 20 ms to reach the 10% level of peak luminances (see also Figure 15 in Chapter 3).

---

[†] These authors tested 33 models; the remaining 12 models were tried out with the same method by van der Heiden and Krüger.

**Table 13** The degrees of oscillation *a* of 45 VDT models belonging to 23 different US and European makes.
*Measured screen surface: 5×7 cm. Ambient illumination: 400 lx.* Preferred luminance of characters.

| Number of models | Degree of oscillation *a* | Ergonomic evaluation |
|---|---|---|
| 8 | 0.02–0.08 | Low degree of oscillation; no risk of flicker. Good condition. |
| 10 | 0.09–0.19 | Moderate degree of oscillation; low risk of flicker. Acceptable. |
| 12 | 0.20–0.39 | High degree of oscillation; flicker possible. Not recommendable. |
| 15 | 0.40–1.0 | Very high degree of oscillation; flicker visible. Intolerable. |

*Ergonomic evaluation of degrees of oscillation*

The ergonomic evaluation in Table 13 is based on the following considerations: degrees of oscillation of VDTs with slow phosphors showed figures similar to those of phase-shifted fluorescent tubes; such low degrees of oscillation do not generate visible flicker or at least have a low risk of flicker and can therefore be ranked as 'good' or 'acceptable'. Degrees of oscillation with figures for *a* of more than 0.2 have short phosphors with very high peaks of luminance. These degrees of oscillation correspond to or exceed those of not-phase-shifted fluorescent tubes: the VDTs showed visible flicker and must be ranked as intolerable.

*Conclusions for flicker-free VDTs*

In conclusion the following recommendation can be deduced for the present. *Preference should be given to CRTs with a degree of oscillation of character luminances comparable to figures shown by phase-shifted fluorescent tubes: a should be lower than 0.2. Refresh rates of 80–100 Hz with phosphor decay times of about 10 ms for the 10% luminance level would correspond to low degrees of oscillation of less than* $a = 0.2$ *and can be recommended.*

*Flicker-free reversed displays*

Since the sensitivity to flicker increases with the size of the oscillating light source, the risk of visible flicker is greater on reversed displays than on screens with bright characters and dark background.

This phenomenon has in fact been observed in several experimental studies (8, 97, 213), where CFF thresholds were measured at refresh rates of about 90 Hz. Therefore the following tentative recommendation seems to be reasonable: *Reversed displays should have refresh rates of 100 Hz associated with rather slow phosphors with decay times of about 10 ms for the 10% luminance level.*

Östberg *et al.* (166) recommend in the guidelines of the

Swedish Telecommunication Administration that displays with a light background should have a minimum refresh rate of 70 Hz and a maximum afterglow time of 100 ms. This minimum refresh rate of 70 Hz seems to be too low and does not correspond to the results of the experimental studies mentioned above.

## 6.5. Sharpness of characters

One important characteristic of image quality is the sharpness of characters or image resolution. In fact, for easy and precise accommodation the characters should have sharp borders. Figure 37 demonstrates display characters with sharp and with blurred borders.

**Figure 37  Sharpness of displayed characters.**
*Left: character with sharp edges. Right: character with large blurred border areas and poor sharpness.*

*Effects of poor sharpness*

It is generally accepted that *display characters with sharp outlines guarantee comfortable legibility with good reading performance, whereas characters with blurred edges cause low visual comfort with reduced performance.*

There are only a few studies on effects of sharpness of VDT characters. For example, Gomer and Bish (62) studied the effects of image resolution on evoked potentials of the brain. (Evoked potentials are indicators in the electroencephalogram of sensory stimuli.) An image of higher resolution produced a stronger and more clearly defined evoked potential than an image of lower resolution. The interpretation of this result is of course very difficult but it indicates that image quality is associated with a neurophysiological response of the brain.

*Sharpness and accommodation*

Korge and Krueger (109) investigated the effects of edge sharpness of characters on accommodation. They observed that blurred characters lead to a shift of accommodation towards the resting position. The authors believe that blurred characters render the precise visual focusing of characters more difficult because of an insufficient sensory control of the accommodation mechanisms. Rupp *et al.* (174) investigated the effects of a sharp and a blurred VDT text on five subjects. Sharpness was determined by measuring the luminance distribution across a vertical stroke with the MTF procedure.

The sharp characters had a blurred border zone of 0.25 mm, whereas the focused characters had one of approximately 0.35 mm. The measurements showed that after a five minute reading task the mean values of accommodation and stability of accommodation were of the same order for both conditions. The results are not very conclusive, since a border zone of 0.35 mm is by no means very poor sharpness; furthermore, the reading time of 5 min must be considered very short for a test of this kind.

*Measuring sharpness*

Printed letters of good quality have sharp edges, whereas VDT characters exhibit a relatively blurred border area. There is no accepted standard procedure at present to measure sharpness of VDT characters.

Howell and Kraft (89) determined the sharpness of printed letters by measuring the blurred border zone in relation to the width of letters. They found a good correlation between a blurred border zone and the sharpness rated by subjects. This was later confirmed by Gould for CRT displays (83).

Snyder and Maddox (187) and Snyder (189) used the MTF procedure to assess the degree of sharpness of displayed characters.

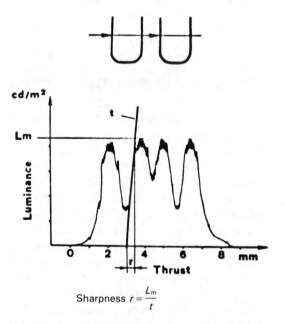

Figure 38 A measuring procedure to determine the blurred border area in mm according to the above mentioned formula.
$L_m$ = *peak luminance in the character bar;* t = *tangent of the slope;* r = *blurred border area in mm. According to Bräuninger et al. (20, 21, 22).*

Bräuninger et al. (20, 22) used the equipment described in Section 6.3 and adopted the following method: the microscope was moved across the capital letters 'U' at a speed of 5 mm/min. The procedure of measuring the blurred border area of these characters is illustrated in Figure 38.

*Sharpness of 45 VDT models*

The tangent of the slope of the increasing luminance in the border zone of the capital letter 'U' is determined and the resulting distance r in mm characterizes the width of the blurred border zone. With this procedure the same 45 VDT models as mentioned in Section 6.4 were tested and evaluated. These results are reported in Table 14.

**Table 14 The character sharpness of 45 models of 23 different VDT makes.**
*Sharpness is expressed as the width of the blurred border area r in mm. The figures are mean values of 9 measurements taken from 9 locations on the screen at preferred luminances. According to Bräuninger et al. (20, 22).*

| Number of models | Width of border area $r$ (mm) | Ergonomic evaluation |
|---|---|---|
| 5 | 0.10–0.29 | No blurred borders. Good sharpness. |
| 17 | 0.30–0.39 | Blurred edges just visible. Unsatisfactory. |
| 12 | 0.40–0.49 | Marked blurred edges. Insufficient sharpness. |
| 7 | > 0.50 | Blurred zone twice as large as that of good sharpness. Unacceptable. |

The ergonomic evaluation is based on the fact that the resolution capacity of the eye for a viewing distance of 60 cm lies at about 0.26 mm. Therefore a character with a blurred zone of less than 0.30 mm appears to have sharp borders. The larger the blurred area the lower the sharpness.

If the luminance of characters is increased above the preferred level, in most models the blurred zones get wider. With maximum luminance all models exhibit poor sharpness.

Poor sharpness is often found to be a fault of the focusing device of the CRT, and in some cases antireflective equipment, such as micromesh filters, substantially reduce the sharpness of characters.

*Conclusions for good sharpness*

Many customers do not have the means to measure sharpness, since the methods that do exist are difficult and costly to use. *The best advice for VDT end-users is to examine character sharpness by eye and to avoid a model with visible blurred edges of characters.*

Manufacturers and selling organizations can be recommended to *choose CRTs and antireflective devices which will guarantee a blurred border zone of characters measuring less than 0.3 mm.*

## 6.6. Character contrasts

Another important photometric factor determining the image quality of displayed information is the luminance contrast between the image and its background.

Some individuals prefer relatively bright characters on the screen together with high character contrasts, whereas others prefer relatively dim characters with low character contrasts. This is why most VDT makes offer a means to adjust the character brightness.

*Contrast ratio*

Various parameters have been produced to describe the luminance difference between characters and screen background. Among them are contrast ratio, MTF and several formulae to define contrast and percent contrast. The easiest figure to be assessed is certainly the contrast ratio: it is the higher luminance divided by the lower luminance. Although the MTF is a more appropriate contrast specification, contrast ratio shall be used here because this parameter is easy to understand and can to some extent be imagined.

Printed texts of good quality usually have high contrast ratios of 20:1 and more. Ratios of 10:1 are considered to be suitable.

Most studies on character contrast are related to the ratio between characters and the screen backgound without a displayed text. Since many VDTs show an increased luminance in the spaces between the characters it seems to be appropriate to take into consideration the contrast ratios between bars of characters and these spaces.

Several studies have been carried out in recent years with the aim of finding out possible influences of character contrasts on legibility of display symbols. The results are summarized below.

*Studies on contrast ratios*

Mourant et al. (146) studied the influence of reducing character contrast with a VDT and with a printed text on two subjects. Contrast ratios of 20:1 were found to be good for the display and 5:1 for the printed text.

Shurtleff (181) determined the legibility of character contrasts and concluded that the minimum contrast ratio acceptable for general display conditions is between 10:1 and 18:1. Furthermore, Shurtleff states that the luminance of characters *per se* should not be lower than 35 cd/m$^2$.

Kokoschka and Bodmann (106) carried out two adjustment experiments and one performance test on 10 subjects to find limits and optimal values of lighting variables at VDT workstations. The authors concluded that character contrast ratios of 10:1 are optimal, while ratios of 2.5:1 must be considered as minimal and ratios of 15:1 as maximal.

*Adjusted contrast ratios in offices*

Adjusted character luminances and the resulting contrast ratios were observed by Läubli et al. (120) in a field study. VDT operators engaged in conversational tasks adjusted the character luminance in such a way as to keep character contrasts in the range between 2:1 and 31:1, the mean contrast ratio being 9:1. It was obvious that factors such as sharpness of character or ambient light had a certain influence on the adjusted luminance of characters. This explains the rather wide range of adjusted contrast ratios.

*Sharpness and contrast ratio*

Snyder and Maddox (188) and Snyder (190) draw attention to the important reciprocal relationship between sharpness and contrast of characters: poor sharpness requires higher character contrast if good legibility is to be maintained. This applies in particular if the contrast ratio between bar luminance and inter-space luminance (space between two characters) is considered. CRTs with an insufficiently focused electron beam disclose increased interspace luminance and the corresponding contrast ratio might be as low as 2:1.

*Adjusted character luminances*

Snyder (190) concludes that any symbol luminance above about 65 $cd/m^2$ is adequate as long as sufficient contrast is maintained. The above mentioned field study (120) revealed that operators engaged in conversational jobs prefer character luminances in the range 9–77 $cd/m^2$ with a median value of 33 $cd/m^2$. These rather low figures might to some extent be due to poor character sharpness.

*Screen background luminances*

The luminance of the screen background depends on the illumination level in the room and on the reflection characteristics of the display screen. It is therefore hardly possible to recommend a precise luminance for the screen background. In the above mentioned study by Läubli (120) the measured background luminances ranged between 1 and 11 $cd/m^2$ with a median figure of 4 $cd/m^2$. This gave a mean contrast ratio of 9:1 for characters. In a Swedish telephone information centre Shahnavaz (180) observed screen background luminances of between 0.2 and 5.6 $cd/m^2$ with a mean luminance of only 1.3 $cd/m^2$ during the day and 2.0 $cd/m^2$ during the night shift.

Some authors and organizations recommend a rather high screen background luminance of between 15 and 20 $cd/m^2$ (28). Such background luminances will mislead operators into adjusting to high character luminances of more than 100 $cd/m^2$ with the risk of obtaining poor sharpness and visible flicker. Thus it is advisable to keep the screen background luminance below 8 $cd/m^2$.

*Character contrasts of 45 VDT models*

Bräuninger et al. (20, 21, 22) studied the character contrasts of 45 VDT models under the standardized lighting conditions mentioned in Section 6.3.

As shown in Figure 38 the luminances recorded with the microscope crossing the bars of the capital letter 'U' do not drop to the background level in the space between letters but remain at a higher level of luminance. This luminance in the space between letters is called the 'rest luminance' and is expressed as a percentage of the luminance measured inside the bars of the letter 'U' according to the following formula:

$$\text{Character contrast} = \frac{L_p}{L_m} 100\%$$

where $L_p$ = rest luminance in the space between letters; and $L_m$ = mean luminance of a dot of 0.1 mm in the middle of a bar.

The character contrasts determined on 45 VDT models according to this procedure are reported in Table 15.

**Table 15 Character contrast of 45 VDT models.**
*The character contrasts are expressed as rest luminance in the space between two 'U's and measured at preferred character luminances between 20 and 50 cd/m² in a dark room.*

| Number of models | Rest luminance (%) | Approximate contrast ratio | Ergonomic evaluation |
|---|---|---|---|
| 19 | < 17 | > 1:6 | Good character contrast. |
| 10 | 17–25 | 1:6–1:4 | Acceptable, with good sharpness. |
| 6 | 26–33 | 1:4–1:3 | Insufficient character contrast. |
| 9 | > 33 | < 1:3 | Very poor condition. |

*Ergonomic evaluation of character contrasts*

The ergonomic evaluation in the table is based on the following considerations:

*A character contrast ratio of 10:1 between bright letters and general dark screen background has become a generally accepted industrial norm for display design.* The space between characters in well-designed displays shows a slightly increased backgound luminance with figures between 8 and 17% of the character luminance. *It is therefore reasonable to accept rest luminances of this order; they are associated with corresponding contrast ratios between 6:1 and 10:1.* Such conditions guarantee good character contrast.

Rest luminances between 17 and 25% correspond to low contrast ratios. But, if the above mentioned relationship between sharpness and contrast of characters is taken into consideration (190), it can be concluded that a sharpness $r$ of less than 0.3 mm combined with a character contrast of about 5:1 should still guarantee fairly good legibility. If one of the two parameters does not meet the recommended level, then the other one at least should show an optimum figure. The best of the studied 45 models did indeed disclose a high sharpness of

*Influences on the rest luminance*

$r = 0.19$ mm and a rest luminance of 8% with a contrast ratio of 12:1.

Bräuninger (21) measured a great increase in rest luminance when the character luminance was shifted from preferred levels to maximum brightness. The mean values of the rest luminance of 27 VDT models increased from 28 to 40%. This observation corresponds to the decrease in sharpness when increasing character brightness. Furthermore, a certain increase in rest luminance was observed when the measurements were carried out in an illuminated room (400 lx).

## 6.7. Stability of characters

If the electron beam is well regulated the typeface appears stable. If the regulation is insufficient the characters show a temporal instability. This phenomenon occurs in the form of drift, jitter or disturbed linearity.

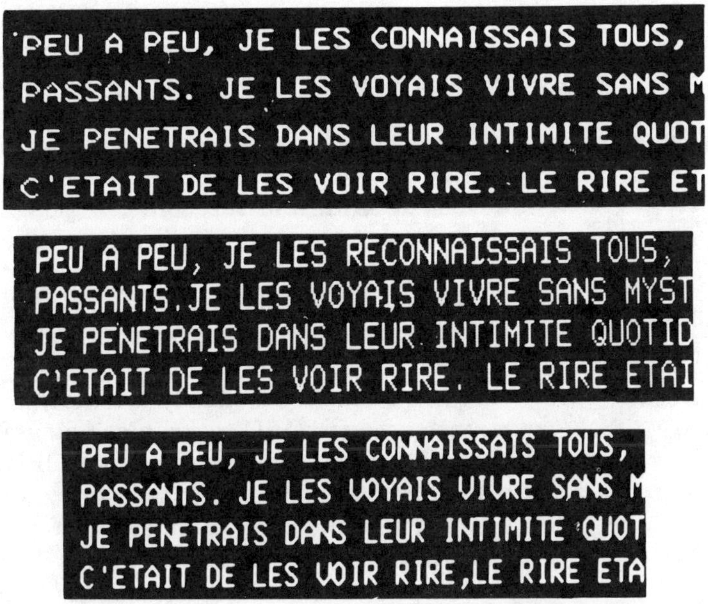

**Figure 39 Three sections of three different VDT screens.**
*Upper: the bars of the letters are sharp and well contrasted to the immediate surroundings. The letters have a good height to width ratio, and the spaces between the letters and between the lines are large.*
*Middle: the height to width ratio is not appropriate. The spaces between the letters and between the lines are too small.*
*Lower: the space between letters is not constant, the bars are often merged. This phenomenon is due to the temporal instability of the characters, mainly caused by the drift of symbols. The whole face is poor.*

*Drift*

Drift is a change in the position of a symbol and can cause a merging of characters, as illustrated in the lower cutting of Figure 39.

The movements of the drift phenomenon are rather slow, of the order of 10–30 s per drift.

*Jitter*

Jitter is a brief, small, abrupt and repetitive change in the position of a symbol. There may be two main reasons for jitter. First, the noise in the electronic line and in the image deflection circuits may cause irregular displacements of single dots. Second, jitter can be produced by AC fields of external sources which may be superimposed on the deflection field. The operators report irregular movements or additional blurring of characters. Bauer (10, 11) studied the phenomenon of jitter in reversed character presentation (dark letters on bright background). For 10 subjects the mean threshold value for jitter movements at 10 Hz was 25.4 $\mu$m, which corresponded to 17.5 seconds of visual angle. Movements with 15 $\mu$m jitter amplitude were just below the threshold for the most sensitive observers. The author concluded that for VDT operators using 80 cd/m$^2$ bright screens the physical jitter at 10 Hz should be less than 15 seconds of arc. At jitter frequencies above 30 Hz movements become blurred. Even if the jitter-induced movements are eliminated, other flicker-like interferences may occur in the border area of the bright background. The subjects reported movements when the eye fixated the border of the screen; moreover, they were aware of flashes appearing in the border region. The author believes that these apparent flashes are due to saccadic eye movements. These phenomena depend on the contrast between screen and border zone as well as on the width of the black frame around the screen.

*Disturbed linearity*

Disturbed linearity is mainly a drifting of lines, appearing on screens used for CAD tasks, block diagrams and other kinds of drawings. It can also be due to brightness fluctuations with a stable position of characters. All types of temporal instability can be superimposed on the displays and the identification of single types may be difficult.

Fellmann et al. (55) and Bräuninger et al. (20, 21, 22) assessed the stability of characters in a simple way: the luminance of a dot of diameter 0.1 mm in the bar of a character was continuously recorded with a microscope. Recorded luminance appeared as a straight line for CRTs with high stability whereas those with low stability revealed characteristic changes of luminance. Figure 40 shows examples of recordings from different VDT makes.

In order to compare and evaluate the stability of characters the recorded variations were determined and their range expressed as variance in % of the maximum luminance.

*Visual strain and photometric characteristics of VDTs* 81

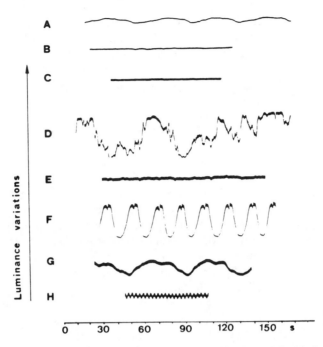

**Figure 40 Recordings of image stability for eight makes of VDTs.**
*The luminance of a dot of 0.1 mm diameter located in the middle of a bar of a character, is continuously recorded. The temporal instability of the characters generates characteristic luminance variations.*

**Table 16 The stability of characters.**
*Expressed as the deviation in % from the maximal luminance of a bar. The figures are mean values of nine measurements recorded at nine locations on the screen with preferred adjustments of luminance.*

| Number of models | Deviation (%) | Ergonomic evaluation |
|---|---|---|
| 8 | < 6 | Very poor stability. |
| 25 | 6–20 | Sufficient stability. Variations hardly perceptible. |
| 5 | 21–40 | Insufficient stability. Variations are annoying. |
| 8 | > 40 | Unacceptable. Characters may merge and reduce readability. |

Temporal instability was therefore determined according to the following formula:

$$S = \frac{L_{max} - L_{min}}{L_{max}} 100\%$$

where $S$ = temporal instability; $L_{max}$ = maximal luminance; and $L_{min}$ = minimal luminance.

It was observed that a variance of less than 5% is not perceived; this figure served as the basis for the evaluation of

46 VDT models, reported in Table 16. There is general agreement that visible temporal instability is annoying for an operator. The legibility of single symbols is usually not reduced (181, 194) but if the characters merge it is very likely that readability becomes impaired.

## 6.8. Reflections on screen surfaces

In Section 5.5 (Appropriate lighting) the reflections which may appear on the glass surfaces of screens were discussed. Reflections are generated by two surfaces: the irregular surface of the phosphor is the source of diffuse or veiling reflections. The polished surface of the front glass is the source of specular reflections. The possibilities of reducing reflections by means of appropriate arrangements of displays, lights and windows were enumerated and illustrated in Figures 26–30.

It is certainly not always possible to completely avoid disturbing bright reflections by these measures. That is why many manufacturers have developed anti-reflective techniques and devices. Some of these systems shall be briefly described and evaluated below.

*Micromesh filters*   Micromesh filters are fine fabrics placed directly onto the glass surface or hung in front of the display screen. Only light falling perpendicularly on the mesh can pass through it and reach the screen. A great part of the light in the room will not fall perpendicularly to pass through the mesh but will be absorbed. This reduces both specular and diffuse reflections. However, micromesh filters also have drawbacks: since the emitted light from the characters is reduced their luminance decreases. This drop in luminance is particularly pronounced if the screen is not viewed head-on. Another drawback is the reduction in sharpness of characters. The lower luminance induces the operator to increase the character luminance which causes even poorer sharpness and higher rest luminances between the characters. Some micromesh filters collect dust, which reduces the character luminance even more.

*Neutral density filters*   Neutral density filters also reduce the amount of light that passes through them. Room light first passes through the filter before it is reflected by the screen surface. The reflected light passes through the filter a second time on its way to the operator. In each process some light gets lost. In comparison the light emitted by the characters only passes through the filter once on its way to the viewer. As a result, the brightness of the characters is reduced less than that of the reflected background and the characters stand out more distinctly. As is true of all filters, it is important that neutral density filters be able to resist the scratching and abrasion of normal use and cleaning which can rapidly degrade character resolution.

*Coloured filters*

Coloured filters function the same way as neutral density filters except that they change the colour of the light that passes through them. The panel of experts of the US Academy of Sciences (153) reports on a study of one neutral and several coloured filters. The contrast and the luminance of VDT characters were measured on screens with smooth and matt surfaces. The measurements were made in total darkness in the presence of a specular source (light box) and a diffuse reflection source (room light and slide projector). All filters reduced the luminance of the display characters. It was concluded that when a filter is used the VDT must be operated at a higher beam current, which causes the phosphor to age more rapidly, thus reducing the lifetime of a CRT. For the matt-finish screen, with specular reflection present none of the filters resulted in a significant improvement over the no-filter condition. For both the smooth and matt finish screens several filters moderately improve the contrast under diffuse reflection. Three coloured filters (amber, grey and green) not only did not improve contrast for either condition but resulted in poorer contrast in several experiments. The panel conclude by saying, "In general, filters are more effective in reducing diffuse reflections than in reducing specular reflections. This is unfortunate because specular reflections cause the greater loss of contrast and probably contribute more to problems encountered in viewing VDTs".

*Etching or roughening the screen glass*

Etching or roughening the front glass surface is an often applied procedure. It is a chemical or mechanical treatment of the outer surface of the front glass in order to produce an irregular surface. By reflecting light in all directions the roughened surface reduces the brightness of reflected images, and renders their borders less distinct. The more the surface is roughened the more the reflection is broken up and diffused. The reflected image becomes softer because it is dispersed over a large area.

The light rays generated by the excited phosphor are also dispersed when they pass through the front glass, which produces a blurring of the display characters. The roughening or etching procedure is moderately efficient in reducing reflections and the effects on character sharpness can be kept slight if the surface is not roughened too much. In fact, several VDT makes with roughened front glass produced reduced reflections but nevertheless good character sharpness.

*The Paci study*

Paci (167) studied the effects of VDTs with a chemically etched screen, a micromesh and a coloured plastic filter and compared them with an untreated, polished CRT. In a legibility test, all the considered anti-reflection devices improved performance over performance with an unprotected CRT. However, although the roughened display gave better performances than the two filters, the eight test subjects preferred the micromesh filters. The results must be interpreted with caution since the glare source used was unusually high and very disturbing.

*Quarter-wave coating*

Quarter-wave coatings are the same type of thin film coating often used on high-quality cameras. The film layer has a thickness of a quarter of the wavelength of light. Reflections from the front surface of the coating interfere with and reduce reflections from the glass surface. Since more room light is transmitted through the face of the CRT and is trapped there, less is reflected to the user. One great advantage of quarter-wave coatings is that they do not diminish the sharpness of characters. On the other hand, these thin film coatings do not reduce the diffuse reflections from the phosphor. Other drawbacks are the sharp outlines of the remaining image reflections and that the surface gets easily soiled by fingerprints.

*The Haubner–Kokoschka study*

Haubner and Kokoschka (80) studied specular and diffuse reflections on VDT screens equipped with a micromesh filter, with dark and bright etching and with quarter-wave coatings. Reflections, visual performance and subjective appraisal were determined for preferred character luminance. The comparison with a control CRT (without any antireflective device) disclosed a marked reduction in diffuse reflection for the micromesh filter and the etching protection. The specular reflection was reduced with all antireflective devices but most of all with the micromesh filter. All protected screens, except that with the micromesh filter, increased legibility performance. The subjects rated the micromesh filter and the two etching protections as "good" and the quarter-wave coating as "acceptable/bad". On the whole the subjects preferred the dark screens with the micromesh filter and dark etching.

*The Habinek study*

The same types of antireflective devices were studied by Habinek et al. (78). Performance in a legibility task revealed no appreciable difference between a micromesh filter, a quarter-wave coating and a screen surface etching. It is obvious that no final conclusions can be drawn from these two studies: the experimental conditions are certainly not identical with the daily work in an office and there are long-term maintenance aspects that were not considered, e.g., dust collection on micromesh and fingerprints on quarter-wave coatings.

*Polarization filters*

Polarization filters polarize the incident light and partially reduce reflections. The best model is the circular polarizer filter with an antireflection coating which reduces both specular and diffuse reflections. The panel of experts of the US Academy of Sciences (153) describe it as follows: "The outside surface of this type of filter is coated with antireflective materials. The rest of the filter package consists of substrate material sandwiched around the more delicate components, a linear polarizer and a quarter-wave plate, which is called the circular polarizer. This element converts unpolarized incident light to circularly polarized light". Because of the optical physics of the circular polarizer, the light is prevented from getting back through the filter in much the same way that light is blocked by crossed linear polarizers. This type of filter reduces specular reflections from the filter itself through the use of antireflective coating and by eliminating specular reflection from the VDT screen through use of the circular polarizer. Diffuse reflections are reduced primarily by the light attenuation effects of the polarizer material, which allows only 35% of the incident unpolarized light to pass to the phosphor surface. The reflected diffuse light is again reduced to about 35% as it passes back through the filter toward the user. The panel of experts (153) say that the circular polarizer filter with the antireflective coating is the most expensive filter available and probably one of the most effective. Bräuninger (21) observed double images on a polarization filter without antireflective coatings — a serious drawback.

Table 17 Reflected luminances generated by a light source of 100 W on screen surfaces.

| Tested antireflective measure | Number of models | Reflected luminance (range) (cd/m$^2$) | Observed drawbacks |
|---|---|---|---|
| Makes without anti-reflective devices | 5 | 525–2450 | Very strong reflections. |
| Quarter-wave coating | 2 | 26–46 | Easily soiled; good reduction of glare. |
| Etching or roughening treatments | 18 | 143–235 | Moderate reduction of sharpness; moderate abatement of glare. |
| Micromesh filters | 7 | 33–72 | Poor sharpness, dark screen background; good reduction of glare. |
| Polarisation filters (without coatings) | 2 | 160–1480 | Easily soiled, double images; slight reduction of glare. |
| Coloured filters | 2 | 205–470 | Poor sharpness; moderate reduction of glare. |

*Reflected glare and drawbacks of 44 VDT models*

Bräuninger et al. (20, 21, 22) compared the reflected luminance of a light source for 44 different VDT models. Five of them had no protection at all and were used to compare the effects of five different antireflective measures. The source of glare was a light source of 100 W placed at a distance of 168 cm from the screen. The incident light formed an angle of 26° with the vertical axis. The reflected luminance was determined with the aid of a Tektronix instrument at the same angle on the opposite side of the vertical axis. The results and comments on observed drawbacks are reported in Table 17.

Since it was not possible to compare the antireflective screens with those without protection the measured reductions in reflected luminance are not very reliable and it is only possible to deduce those evident tendencies which are indicated in Table 17.

*Conclusion*

*All antireflective techniques have serious drawbacks. If efficiency is weighed against drawbacks, the quarter-wave coatings and the etching/roughening procedures are preferable, whereas micromesh, polarization and coloured filters cannot be recommended. The adequate positioning of lights and appropriate positioning of the screen with respect to windows remain the most efficient preventive measures.*

## 6.9. Size of characters and typeface

The main geometric elements of characters are:

Height
Width
Height:width ratio
Stroke width
Font
Horizontal spacing
Vertical spacing (space between lines)

These elements are of great importance for easy identification and reading as well as for visual comfort.

The requirements for printed texts are well known and described in all ergonomic textbooks.

*Character size*

Since the resolution of VDTs is usually poorer than that of printed materials, character size and character contrast call for special attention. Shurtleff (181), Snyder and Maddox (188, 190), Vartabedian (202) and Stewart (194) have conducted the most detailed studies in this field, the main parameters being identification accuracy (legibility) and in some cases readability.

In general, character size is defined as the physical height of upper case letters. The usual measurement of a character is from top to bottom edge. If the edge is not sharp, it is recom-

**Table 18 Recommended geometric elements of characters.**

| For viewing distances of 50–70 cm | Recommended figure |
|---|---|
| Height of capital letters (mm) | 3–4.3 |
| Width of capital letters (% of height) | 75 |
| Stroke width (% of height) | 20 |
| Distance between characters (% of height) (Two dots are recommended) | 25 |
| Space between lines (mm) | 4–6 |
|   as % of height of capital letters | 100–150 |

mended to use the 50% luminance criterion, that is the mean value between the bright and dark areas.

There is general agreement among the authors as to the size of characters; provided that suitable contrast ratios and character luminances are given *the range for appropriate character sizes is 16–25 minutes of visual angle. Thus a suitable character height is 3 mm at a viewing distance of 50 cm and 4.3 mm at a viewing distance of 70 cm (corresponding to 20 minutes of visual angle).*

Larger sizes cause dot dissociation, rendering legibility more difficult. Furthermore, too large and too distant characters are associated with a lower number of letters in the visual reading field, causing lower readability (see Section 4.7).

Larger characters should only be used when recognition of each single letter is important, such as for tasks that involve searching or typing from material displayed on the screen.

All generally recommended geometric sizes of VDT characters and of spaces are given in Table 18.

*Geometric elements of 18 VDT makes*

Bräuninger (21) determined some geometric elements of characters of 18 different VDT makes. The results are reported in Table 19.

**Table 19 Geometric elements of displayed characters of 18 different VDT makes.**
*All distances (in mm) were measured from dot centre to dot centre.*

| Geometric element | Recommended figure | Mean value | Range | Remarks |
|---|---|---|---|---|
| Height of capital letters (mm) | 3–4.3 | 3.3 | 2.5–4.3 | 4 are below 3.0 |
| Width of capital letters (mm) | 2.3–3.2 | 1.8 | 1.1–2.2 | 13 are below 2.3 |
| Stroke width* (mm) | 0.6–0.8 | 0.6 | 0.3–0.8 | 8 are below 0.6 |
| Distance between characters (mm) | 0.7–1.1 | 1.0 | 0.7–1.4 | all within recommended range |
| Distance between lines (mm) | 4–6 | 3.5 | 1.5–9.6 | 14 below 4.0 |

*Stroke width was measured between the 50th percentile of border luminance of bars.

The most frequent faults are certainly insufficient space between lines and narrow width of characters. These characteristics are illustrated in the middle and lower panels of Figure 39 (p.79). The reason for these shortcomings is most probably the attempt of the designers to get 24 lines onto a 12 inch screen and to fill each line with as much text as possible. The results show that in practice the number of lines and words displayed is often increased at the expense of line distance and character width. There is a conflict of aims: for word processing it is useful to have a full page of information; for data retrieval tasks, on the other hand, larger characters and fewer lines of text would be a great advantage.

*Dot matrix*

In Chapter 3 it was pointed out that the spaces between dots should not be visible and that a dot matrix of $9 \times 11$ offers better legibility than a dot matrix of $7 \times 9$ or $5 \times 7$ (187). The experts of the Bell Telephone Laboratories (13) state that the smallest acceptable size of a dot matrix is $5 \times 7$, although $7 \times 9$ or $9 \times 11$ are generally preferred.

The 18 VDT makes studied by Bräuninger (21) exhibited the following dot matrix figures:

8 makes: $7 \times 9$
7 makes: $5 \times 7$
1 make: $7 \times 7$
1 make: $6 \times 9$
1 make: $4 \times 7$

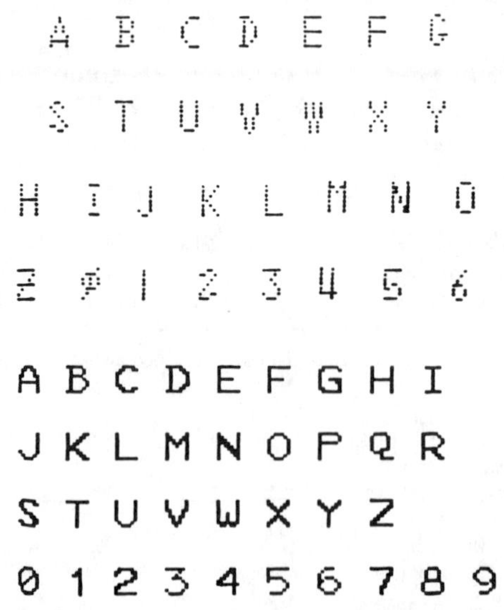

A B C D E F G
S T U V W X Y
H I J K L M N O
= ? 1 2 3 4 5 6

A B C D E F G H I
J K L M N O P Q R
S T U V W X Y Z
0 1 2 3 4 5 6 7 8 9

**Figure 41** Parts of a $5 \times 7$ matrix font (above) and a $7 \times 9$ matrix font (below).
*The risk of visible separation of the dots is greater with the $5 \times 7$ than with the $7 \times 9$ matrix:. Above: Lincoln-Mitre font. Below: Huddleston font.*

Figure 41 shows a 5×7 matrix with the Lincoln-Mitre font and a 7×9 matrix with the Huddleston font.

From these studies and from general experience it is concluded that for capital letters dot matrices of 9×11 and of 7×9 are to be recommended, but dot matrices of 5×7 or 6×9 are still sufficient.

*Small letters*

In the relevant literature no reference is made to recommended sizes for small letters. The mean height of small letters was 2.4 mm the mean width 1.6 mm for 25 different VDT makes examined. It is reasonable to recommend the height of small letters to be about 70% of the height of capital letters and the width about 60% of the height of capital letters.

*Interline distance*

In Section 4.7 it was pointed out that the distance between lines is of great consequence for correct and precise line saccades. Bouma (18) has indeed observed that an interline distance equal to the height of two lower case characters reduces the visual reading field to seven letters. He recommends a minimum interline distance of about 1/30 of the line length; that means that for a line length of 15 cm the interline distance should be 5 mm.

*Upper and lower case letters*

All experts agree that upper case characters increase readability and accuracy. On the other hand, texts with both upper and lower case characters are generally easier to read than texts with upper case characters only. Both alternatives should be available.

*Descenders*

The legs of lower case characters such as 'g' or 'y' that normally extend below the line are called 'descenders'. The human factors experts of the Bell Telephone Laboratories (13) point out that characters with descenders below the line are easier to read than characters in which the decenders are located above the line.

*Fonts*

The dot matrix does not allow round and differentiated shapes of alphanumeric symbols. Thus the fonts used on CRT displays are rather angular in shape. In most fonts there are letters which resemble others enough to cause occasional confusion. The symbols 'Y' and 'V', '4' and '1', '7' and '1', 'Z' and '2' or '5' and 'S' are often confused. The risk of confusion is less relevant when a normal text must be read than in certain data retrieval tasks.

There are two main requirements with regard to display fonts: *the symbols should have shapes that are easily distinguished from each other and acceptable as reasonable representations of the symbols concerned.*

*Different fonts are available*

Among the fonts proposed for VDTs the following seem to be the most common: Lincoln-Mitre, Huddleston, Hazeltine and IBM 029. Examples of the Lincoln-Mitre and Huddleston fonts are shown in Figure 41, and a part of the IBM 029 font is seen in the upper panel of Figure 39. Shurtleff (181) explored these fonts and arrived at the conclusion that the Lincoln-Mitre font can be recommended since experiments had revealed minimal intrasymbol confusions. Snyder and Maddox (187, 190) and Abramson *et al.* (1) observed significant effects of fonts on legibility and readability. These studies indicated the superiority of the Huddleston font, particularly for flat panel displays which are subject to failures of single picture elements (pixels) or entire lines of addressable pixels. It was demonstrated that the Huddleston font produces shorter text reading times for all levels of percent pixel failure and for all types of display failure. Moreover, the interaction of font and matrix size had a substantial effect on legibility and readability; on a matrix of $5 \times 7$ the legibility of the Huddleston font was superior to that of three other fonts. However, the Huddleston and the Lincoln-Mitre fonts were equally legible on $7 \times 9$ or $9 \times 11$ matrix sizes (187, 190). It is obvious that the matrix size is as important as the shape of characters. This is already well illustrated in Figure 41. The experts of the Bell Telephone Laboratories (13) criticize that the characters '4' and '1' are often confused in the Huddleston font, whereas 'Z' and '2' are confused in the Lincoln-Mitre font.

*The Van Nes font for teletexts*

Recently Van Nes (157, 158) has conducted experiments to design and test a complete set of optimally discriminative alphanumeric characters and punctuation marks for use in teletext. The resulting font is shown in Figure 42.

The font is based on a $10 \times 12$ dot matrix and also includes lower case letters. To increase the discrimination between numerals and capital letters the numeral strokes have a width of 3 matrix elements compared with 2 for the capital letters. This improves the distinction between, e.g., '5' and 'S'. Such difference in stroke width is not used in the Huddleston and Lincoln-Mitre fonts. Although the Van Nes font is designed for teletexts, it seems that its advantages could also benefit VDTs.

*Conclusion*

It seems clear that the Huddleston font, as well as the Lincoln-Mitre and the IBM 029 font, guarantee good legibility and readability. Nevertheless these fonts could still be improved if some of the characteristics of the Van Nes font for teletext were taken into consideration. It must be pointed out that dot matrix sizes often contribute more to the differentiation and legibility of different alphanumeric symbols than the shape of the letters.

**Figure 42 The basic set of the font of the Institute of Perception Research in Eindhoven (the Netherlands).**
*A $10 \times 12$ dot matrix is used for this teletext font. According to Van Nes (157, 158). © F. Van Nes 1983.*

## 6.10. Dark versus bright characters

For the two image polarities the following expressions will be used here:

Bright characters on a dark screen: positive presentation of characters or positive-contrast display.
Dark characters on a bright screen: negative presentation of characters, negative-contrast display or simply reversed display.

*Contrast ratios*

Chapter 3 gave a brief description of the reversed display. In Section 5.3 it was pointed out that reversed displays have background luminances of between 50 and 100 $cd/m^2$ and that such screens do not have the problem of excessive contrast ratios between screen and source document.

The reversed display offers a reading condition which is very similar to conventional dark letters printed on a light paper background. One can therefore assume that this is not only the conventional way of reading but that when used on VDTs it also allows a good general illumination level in the office and thus guarantees good legibility of all kinds of source documents. However, there are several other aspects and potential drawbacks of reversed displays which must also be taken into account.

*Stroke width*

Some VDTs can display both dark and bright characters. These terminals simply invert the signal to the electron gun: any dot that would appear dark is made to glow and any dot that would glow is left dark. The result is a smaller stroke width on dark characters than on bright ones, so the bright dots that delimit a dark stroke on either side will tend to impinge on the stroke itself. The characters may then appear very thin and it is likely that legibility and readability of such dark characters will be reduced.

*The Bauer study on stroke width*

Bauer (11) and Bauer and Cavonius (9) did a thorough study of the problems of stroke width on a reversed display. The 27 subjects had to compare the stroke width of dark and bright strokes. A refresh rate of 85 Hz was associated with a rather fast phosphor similar to $P_4$. For the reversed display a background brightness of 100 cd/m$^2$ was adopted. The results revealed an interesting perceptual effect: observers judge the sharpness of black strokes on a bright background to be substantially higher than that of equivalent bright strokes on a dark background. The subjects preferred on average a black stroke width of 0.38 mm; the stroke appeared grey when the width decreased to 0.3 mm and unsharp when the width reached a figure of 0.42 mm. Dark strokes with a width of 2.64 minutes of arc were perceived as equally large as bright strokes with a width of 2.14 minutes of arc. This means that for strokes to be perceived as equal in width dark strokes must be 23% wider than bright strokes.

*For dark characters it is therefore recommended to select a font that has been adapted to the requirements of negative image presentation, for instance by using up to two dark dots per stroke, in order to guarantee sufficient stroke width.*

*Visible flicker*

Most reversed displays have another serious disadvantage: since the sensitivity to flicker increases with the size of the oscillating source, the risk of visible flicker is greater than in displays with bright characters.

As mentioned in Section 6.4, the range of flicker perception (critical fusion frequency) is higher with reversed displays (between 55 and 87 Hz) than with bright characters on a dark screen (8, 10). Isensee and Bennett (97) as well as van der Zee and van der Meulen (213) also observed a higher sensitivity to flicker with reversed displays than with positive-contrast displays. It is concluded that *reversed displays should have refresh rates of 100 Hz and rather slow phosphors with decay times of about 10 ms to the 10% luminance level.*

*Reflections are less annoying on bright screens*

One can expect that reflected glare would be less troublesome on reversed displays; indeed, dark characters should guarantee a sharper contrast with bright reflections. Bauer *et al.* (7) investigated the effect of specular reflections on the legibility of dark and bright characters. A black and white checkerboard pattern was reflected from a plate of glass placed before the screen. The VDT was operated with a refresh rate of 80 Hz. With the positive-contrast display the letters had a luminance of 120 cd/m$^2$ and the background less than 10 cd/m$^2$; with the reversed display the letters had a luminance of less than 10 cd/m$^2$ and the background 120 cd/m$^2$. The eight subjects had to count the number of capital 'A's in the displayed text and to adjust the reflected luminance of the checkerboard to levels of acceptance or annoyance. In every experimental condition improved legibility was achieved with the bright screen background. In other words: when the screen luminance was higher the luminance of the reflected checkerboard had to be increased in order to reach the same effect. The results were most striking with a low character density on the screen: on average the checkerboard luminance had to be increased more than fivefold to be visible on the bright screen. The authors concluded that adverse effects of reflections are substantially reduced on reversed displays with dark letters and bright screens.

*Performance and preference with reversed displays*

Radl (170) and Bauer and Cavonius (6) observed higher task performances with reversed displays and the operators evaluated dark letters and bright screens as more comfortable and preferable. However, in a follow-up study Cavonius (32) could not confirm the previous results related to performance. McVey et al. (204) studied the effects of different anti-glare devices for positive and negative image displays under adverse lighting conditions on 24 typists. Their task was to key the displayed characters (in four-letter groups). Speed and accuracy were recorded and sessions alternated between dark character and light character presentations. Preference data were obtained by asking each typist whether she preferred the light or the dark characters. Fluorescent lights were arranged behind the typists to produce reflected glare on the displayed image. The ambient luminance on the horizontal table was 1076 lx(!). The display was inclined backwards with an angle of 15° to the vertical and so received an incident light of 861 lx, producing a strong source of glare on the screen.

With high ambient illumination and adverse glare conditions a slightly higher rate of random character recognition occurred with dark than with bright letters. The polarity preference conclusion was that 17 typists favoured light characters whereas 7 favoured dark characters, but this ratio was statistically not significant. Unfortunately the authors gave no information on the photometric characteristics, nor on refresh rate or stroke width of the dark letters. But they point out that "it should be recognized that the effective stroke width of the character font design should be greater with dark character polarity".

*What do we know at the present?*

Up to now no definite conclusions can be drawn from the experiments so far completed. The contradictory results might be due to different photometric characteristics of the tested reversed displays, to different environmental lighting conditions (e.g., glare sources or normal lighting conditions) and to different test procedures to assess legibility. Furthermore, it must be pointed out that as yet no field studies have been carried out and since reversed displays are not yet widely used there is little common experience with such VDTs. However, the facts known about reversed VDTs lead to the following assumptions:

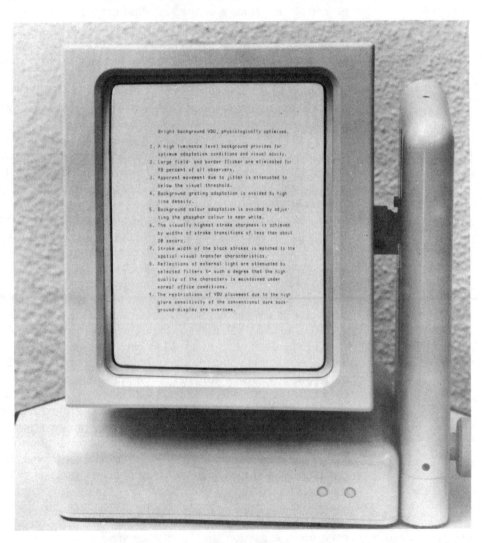

**Figure 43 The experimental unit of a reversed display according to Bauer (11).**
Refresh rate = 86 Hz; phosphor similar to $P_4$; stroke width of characters about 0.38 mm; jitter and flicker are not visible; high sharpness of character edges; background luminance is adjustable.

*Advantages versus drawbacks*

1. Theoretically the bright screen background should have the advantage of reducing the luminance contrast ratio between screen and source document.
2. Reflections on the glass surface of screens should be less disturbing on displays with dark characters and bright background.
3. The general illumination level of the office can be increased to guarantee a good legibility of source documents without creating high contrast ratios with the screen.

But at present reversed displays often exhibit two kinds of drawbacks:

1. With a bright screen there is a definitely greater risk of visible flicker than with positive character presentation. In this respect some studies offered clear results which indicate that reversed displays should have refresh rates of 100 Hz and rather slow phosphors.
2. In most reversed displays the stroke width is too thin, which might decrease legibility and readability. A font with two dots stroke width should prevent this.

It is concluded that the potential advantages of reversed displays are certainly important enough to deserve more research, including field studies with reversed displays free of the above mentioned drawbacks.

*A reversed display free of drawbacks*

Indeed, Bauer (11) demonstrated a reversed display without drawbacks. This experimental unit of a reversed display is shown in Figure 43. It has a refresh rate of 86 Hz and a phosphor similar to $P_4$. The stroke width is about 0.38 mm and the edges are clear-cut, avoiding the appearance of blurred border areas. Its character sharpness is significantly better than that of equivalent bright characters on dark background. The author concludes that such a reversed display closely resembles a printed text on white paper.

# 7. Ergonomic Design of VDT Workstations

## 7.1. Constrained postures are long-lasting static efforts for the muscles involved

*Dynamic versus static work*

There are two kinds of muscular effort: *dynamic effort* and *static effort*.

Figure 44 illustrates these two kinds of muscular activity. The example of dynamic effort is cranking a wheel, the static one, supporting a weight at arm's length.

*Dynamic muscular work is characterized by a rhythmic alternation between contraction and relaxation of muscles.*

*Static muscular work is characterized by a prolonged state of contraction which usually implies a postural stance or constrained postures.* During static effort the muscle is not allowed to relax, but remains in a state of heightened tension. With static work no outward performance is visible — the comparison with an electromagnet suggests itself here. An electromagnet has a steady consumption of energy while it is

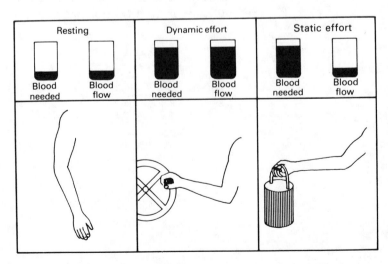

Figure 44   Diagram of dynamic and static muscular effort.

*Blood supply of working muscles*

supporting a given weight but does not appear to be doing visible work.

There is a fundamental difference between static and dynamic muscular effort: during static work the blood vessels are compressed by the internal tension of muscle tissue, so that blood no longer flows through the muscle.

During dynamic work, on the other hand, the muscles act as a pump to the blood circulation. Compression squeezes blood out of the muscles and the following relaxation releases a fresh flow of blood into it. By this means the blood supply becomes several times higher than normal: in fact, the muscle may receive between 10 and 20 times as much blood as it does when resting. A muscle performing dynamic work is therefore flushed with blood and is continuously supplied with high-energy sugar as well as oxygen, while at the same time waste products are removed. That is why dynamic effort can be carried on for a very long time without fatigue, provided a suitable rhythm is chosen. Indeed, walking and to some extent even running can be sustained for many hours without energy deficit in the muscles. There is also one muscle that is able to work dynamically throughout our whole life, without interruption and without tiring: the heart muscle. The upper section of Figure 45 shows the effects of dynamic work on blood supply.

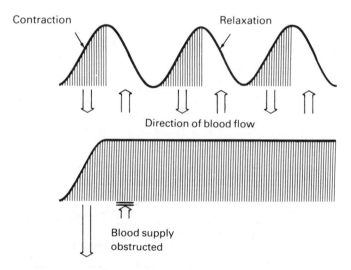

**Figure 45 Blood flow through muscles during dynamic and static efforts.**
*The curves show the variation of muscular tension (internal pressure). Upper: dynamic work operates like a pumping mechanism which ensures the flow of blood through the muscles. Lower: static effort obstructs the supply of blood, energy and oxygen.*

## Ergonomics in Computerized Offices

*Static work throttles blood supply*

In contrast to dynamic work, a muscle that is performing hard static work does not receive sugar or oxygen from the blood but must depend on its own reserves. This is illustrated in the lower section of Figure 45. Moreover — and this is by far the most serious disadvantage — waste products are not being removed. Quite the reverse: waste products, such as carbon dioxide or lactic acid, are accumulating and produce the acute pain of muscular fatigue, often called localized fatigue.

For this reason it is hardly possible to continue static muscular effort for a very long time; the pain will compel one to desist.

*How long can static work be sustained?*

During static effort the flow of blood is reduced in proportion to the force exerted by the muscle. If the effort reaches 60% of the maximum, the flow will be almost completely interrupted; but during lesser efforts the blood circulation will get started again because of decreasing tension in the muscles. When the effort is less than 15–20% of the maximum, the blood flow becomes nearly normal. Figure 46 shows the maximum duration of static contraction in relation to the force exerted.

It seems that static effort which exerts 50% of the maximum force can last no longer than one minute, whereas if the force expended is less than 20% of the maximum the muscular contraction can continue for much longer periods. *General experience leads to the conclusion that static work can be sustained for several hours without symptoms of fatigue if the*

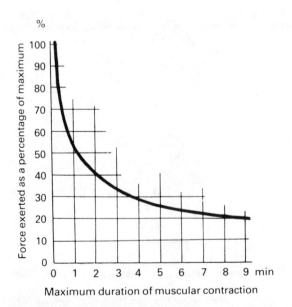

**Figure 46 Maximum duration of static muscular effort in relation to the force exerted.**
*According to Monod (144).*

*exerted force does not exceed about 8% of the maximum force of the involved muscles.*

*Examples of static effort*

In everyday life the human body must often perform static work. Thus, when it stands upright, a whole series of muscle groups in the legs, the hips, the back and neck are stressed for long periods. It is thanks to these static postures that selected parts of the body can be held in any desired position. In the sitting posture the static work of the legs is relieved but a certain amount of static work is still required to maintain an erect position of trunk and head. When lying down, nearly all static muscular effort is suspended; that is why a recumbent posture is the most restful.

*Combination of dynamic and static efforts*

In many cases no sharp distinction between dynamic and static effort can be made. A particular task can be partly static and partly dynamic. Keyboard operating is an example of a combination of both types of muscular work: the shoulders and arms do mainly static work when holding the hands in the typing position, while the fingers perform mainly dynamic work when operating the keys. Since static effort is much more strenuous than dynamic work, the static component of combined effort assumes greater importance. There is a static component in almost every form of physical work.

*Examples of static work*

In the following some of the common examples of tiring static effort are given:

1. Jobs which involve bending the back either forwards or sideways.
2. Holding things, such as books, tools or other loads, with the arms.
3. Operations which require the arms to be lifted up horizontally.
4. Standing in the same place for long periods.
5. Bending the head strongly downwards or upwards.
6. Lifting the shoulders for long periods.

*Constrained postures are certainly the most frequent form of static muscular work.* The main causes of constrained posture are carrying the trunk, head or limbs in unnatural positions. As already mentioned in Chapter 1 the introduction of VDTs leads to an integration of employees in a man—machine system. One of the consequences of this is a restriction of space for physical activities and the origin of constrained postures.

*Localized fatigue and musculo-skeletal troubles*

As already explained, even moderate static work might produce troublesome localized fatigue in the muscles involved, which can build up to intolerable pain. If the static load is repeated daily over a long period, more or less permanent aches will appear in the limbs and may involve not only the muscles but also the joints, tendons and other tissues. Thus long-lasting and daily repeated static efforts can lead to damaged joints, ligaments and tendons. All these acute and chronic impairments are usually summarized under the term 'musculoskeletal troubles'.

Several field studies as well as general experience have shown that such static loads are associated with a higher risk of:

1. Inflammation of the joints (arthritis).
2. Inflammation of the tendon sheaths (tendinitis or peritendinitis).
3. Inflammation of the attachment points of tendons.
4. Symptoms of chronic degeneration of the joints (chronic arthroses).
5. Painful induration of muscles.
6. Intervertebral disc troubles.

*Persistent musculoskeletal troubles*

These symptoms of overstress can be divided into two groups: *reversible* and *persistent musculoskeletal troubles*.

*Reversible symptoms* are short-lived. The pains are mostly localized to the muscles and tendons, and disappear as soon as the static load is relieved. *These troubles are the pains of weariness.*

*Persistent troubles* are also localized to strained muscles and tendons, but they affect the joints and adjacent tissues as well. The pains do not disappear when the work stops, but continue. *These persistent pains are attributable to inflammatory and degenerative processes in the overloaded tissues.* Elderly employees are more prone to such persistent troubles. According to van Wely (208), persistent musculoskeletal troubles are commonly observed among operators who work all the year round at the same machine at which the manual controls are either too high or too low.

*Musculoskeletal troubles related to workstation*

Persistent musculoskeletal troubles, if allowed to continue over a number of years, may get worse and lead to chronic inflammations of tendon sheaths or even deformations of joints. van Wely (208) reported a study of the workplaces of 50 employees of the Philips Company at Eindhoven, who were being treated at the factory clinic for musculoskeletal troubles such as those described above. In 39 out of 50 cases he showed a clear link with bad working postures. At 40 out of 50 workplaces there was an unsuitable layout, which led to constrained postures: in 19 of these the machine was at fault, in 21 it was bad seating.

## 7.2. Body size and the design of workstations for traditional office jobs

*Anthropometric aspects*

In order to avoid constrained postures and to guarantee easy control of machines the design of the workstation must be, among others, adapted to several elements of body size. Here we are soon up against a problem: the great variation in body size among individuals, between the two sexes and different ethnic groups. The science that deals with measuring size, weight and proportions of the human body, is anthropometry. The results of anthropometric research allow us to predict human reach and space requirements and to assess principles for the design of suitable dimensions of workstations.

Several ergonomic textbooks (34, 70, 145, 147) and special publications (3, 45, 66, 111, 167b, 201, 210) make anthropometric data available for the design of workstations, tools and machines. Here only the data relevant to sedentary office work will be discussed briefly.

A preliminary remark is necessary: the differences of anthropometric data between ethnic groups might be important, but the differences between the industrial nations — USA, Germany, France and Great Britain — are so slight that they can be neglected. For instance, the range of mean figures of adult males belonging to these four population groups are for:

Body length:            170 – 173 cm
Back to front of knee:  59 – 60 cm
Top of knee to floor:   54 – 55 cm
Elbow level to floor:   105 – 107 cm

Selected anthropometric data for the design of office workstations are given in Table 20. They were assessed on 15 700 men and 17 700 women in the Federal Republic of Germany. The measurements were carried out on the unclothed body

Table 20 Selected anthropometric data relevant to the design of seated office work.
$\bar{x}$ are mean values in cm. 90% ile = range of 90% of the corresponding population. According to Kroemer (111).

| Part of body | Men $\bar{x}$ (cm) | Men 90% ile | Women $\bar{x}$ (cm) | Women 90% ile |
|---|---|---|---|---|
| Body length | 172 | 160–184 | 161 | 150–172 |
| Top of head above seat | 90 | 84–96 | 85 | 79–91 |
| Eye level above seat | 79 | 73–85 | 74 | 68–80 |
| Elbow height above seat | 24 | 20–28 | 24 | 20–28 |
| Forward reach with grasping fingers | 82 | 75–87 | 70 | 63–77 |
| Back to hollow of knee | 50 | 46–54 | 46 | 43–50 |
| Floor to top of knee | 55 | 51–59 | 50 | 46–54 |
| Buttocks to front of knee | 59 | 54–64 | 57 | 52–62 |
| Thickness of thigh | 14 | 12–17 | 14 | 12–17 |

without shoes. For persons wearing shoes 2.5 cm should be added to men's heights and 4 cm to women's.

*Recommendations are compromises*

The ergonomic recommendations for the dimensions of workstations are only to some extent based on anthropometric data; behavioural patterns of employees and specific requirements of the work itself must be considered too. Thus the recommended dimensions given in textbooks or in various standard works are compromise solutions which may often be quite arbitrary. Another critical remark is necessary: most standard specifications for ergonomic workstations were worked out by committees, in which ergonomics, economics, industry as well as unions or employers were represented. The resulting recommendations seem reasonable and suitable in most cases, but they were seldom seriously tested under practical conditions. It is therefore not surprising when field studies of practical experience do not always confirm recommended standard dimensions.

*Working height*

Working height is of critical importance in the design of working places. If the working height is too high the shoulders or the upper arms have to be lifted to compensate, which may lead to painful symptoms and cramps at the level of neck and shoulders. If, on the other hand, the working height is too low, the back must be excessively bowed, which may cause backache. *Hence the work table must be of such a height as to suit the body length and the activity of the operator. This also applies to sedentary work in offices.* A few studies in this field demonstrate the adverse effects of inappropriate desk

*How office employees sit*

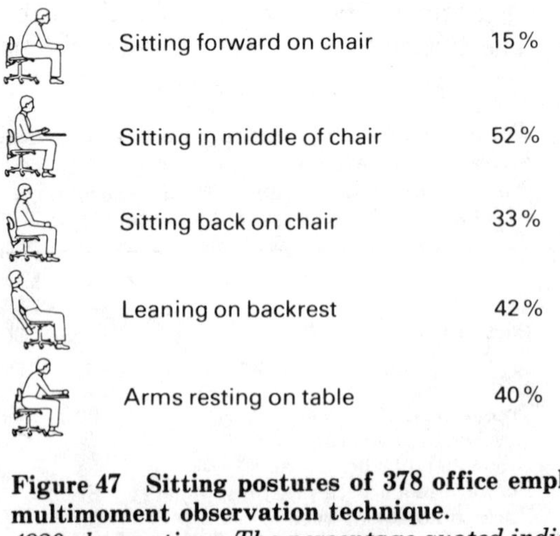

| | | |
|---|---|---|
| | Sitting forward on chair | 15% |
| | Sitting in middle of chair | 52% |
| | Sitting back on chair | 33% |
| | Leaning on backrest | 42% |
| | Arms resting on table | 40% |

**Figure 47 Sitting postures of 378 office employees, as shown by a multimoment observation technique.**
*4920 observations. The percentage quoted indicates how much of the working period was spent in that posture. The two lower observations were seen simultaneously with the three upper postures, which is why the sum of all five characteristics exceeds 100% (67).*

# Ergonomic design of VDT workstations

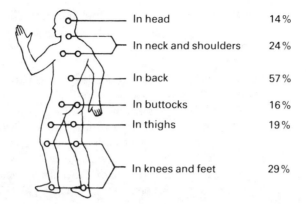

**Figure 48 Incidence of bodily aches among 246 employees engaged in traditional sedentary office jobs.**
*Multiple answers were possible.*

heights: as early as 1962, a survey carried out by Grandjean and Burandt (67) on 261 men and 117 women engaged in traditional office work revealed interesting links between desk heights and musculoskeletal troubles. The work-sampling analysis gave particulars about the different sitting postures, shown in Figure 47.

An upright trunk posture was observed only about 50% of the time, with the trunk leaning against the back-rest about 40% of the time, although most of the chairs were provided with rather poor back supports.

Figure 48 presents the results of the survey on musculoskeletal complaints.

*Links between desk height, sitting behaviour and pains*

The principal anthropometric data and desk heights were assessed and compared with the reports on muscoloskeletal troubles. From a large number of results and calculated correlations, the following conclusions emerged:

1. 24% reported pains in neck and shoulders which most of the subjects, especially the typists, blamed on a too high desk top.
2. 29% reported pains in the knees and feet, most of them small people who had to sit on the front edge of their chair, probably because they had no footrests.
3. *A desk top height of 74–78 cm gave the employees most scope for adaptation to suit themselves*, provided that a fully adjustable seat and footrests were available.
4. *Regardless of their body length, the great majority of the workers preferred the seat to be 27–30 cm below the desk top. This seems to permit a natural position of the trunk, obviously a point of first priority with these employees.*
5. The incidence of backache (57%) and the frequent use of the

*Electromyography of shoulder muscles*

backrest (42% of the time) indicate the need to relax the back muscles periodically and may be quoted as evidence of the importance of a well-constructed backrest.

Several authors have recorded the electrical activity of the shoulder muscles while the subjects worked with different desk heights. It should be remembered here that the electrical activity of a muscle is an indicator of the exerted muscular force; the procedure is called electromyography. Already in 1951, Lundervold (135) investigated the electrical activity of shoulder and arm muscles of subjects operating typewriters at high and lower levels. (At that time electrical impulses were

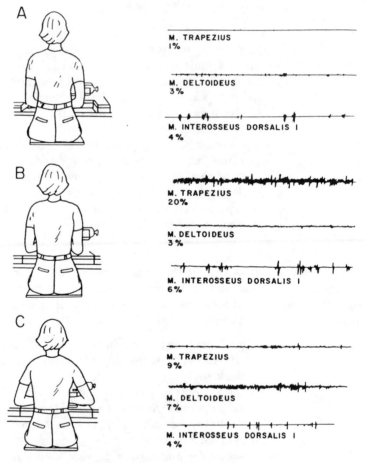

**Figure 49  Electromyographic recording of shoulder muscle activity.** *The figures refer to the percentage of time in the maximum voluntary contraction position. A = optimal height of the typewriter (i.e., home row at elbow height). B = too high, resulting in elevation of shoulders by the trapezius muscle. C = too high, compensated by a sideward elevation of the upper arms by the deltoid muscle. According to Hagberg (79).*

obtained with ink-writer records). The author concluded: "the smallest number of action potentials (electrical impulses) were recorded when the person undergoing the experiment was sitting in a relaxed and well balanced state of equilibrium, or was using a backrest". High-level typewriting was associated with raised shoulders and a strongly increased electrical activity of the trapezius and deltoid muscles. (The trapezius muscle lifts the shoulders, the deltoid the upper arms). Recently Hagberg (79) made a quantitative analysis of the electromyograms of shoulder and arm muscles when typing at different heights. These results are shown in Figure 49.

*Lifting the shoulders is strenuous static work*

Thus a too high working level can be compensated either by lifting the shoulders through contraction of the trapezius muscles or by lifting the upper arms with the deltoid muscles. Moreover, the contraction force of the shoulder-lifting muscle reaches 20% of the maximum force, which would certainly suffice to eventually generate great pains in the shoulder muscles.

There was a classical study carried out in 1951 by Ellis (52) which is often advanced when working heights are discussed. Ellis was able to confirm an old empirical rule: *the maximum speed of operation for manual jobs carried out in front of the body is achieved by keeping the elbows down at the sides and the arms bent at right angles.* This is a generally accepted basis for the assessment of working heights.

*Most important: seat to desk distance*

For sedentary office jobs, working heights must also take note of the optimum visual distance. In some cases the working surface must be raised until the operator can see clearly without having to bend the neck too much. A slight forward stoop, with the arms on the desk, is only minimally tiring when reading or writing, but in order to relax the back *the distance from seat surface to desk top must be between 27 and 30 cm.* As mentioned above, employees sitting at an office desk first of all look for a comfortable and relaxed position of the trunk and often accept a seat height that is bad for legs or buttocks, rather than sacrifice a comfortable trunk posture.

The height of tables which are not adjustable is primarily based on average body measurements and makes no allowance for individual variation. Therefore all table heights recommended are too high for short people, who will need some kind of footrest. On the other hand, tall people will have to bend the neck over the work table which will cause musculoskeletal troubles in the neck and back. Table heights for traditional office work (with the exception of typing!) thus follow the rule that it is more practical to choose a height to suit the tall rather than the short person; the latter can always be given a footrest so that he/she can raise the seat to a suitable level. On the other hand, a tall person given a table

*Conclusion for office desk without typewriter*

that is too low can do nothing about it except fix the seat so low that it might be uncomfortable for the legs.

*Office desks to be used without a typewriter should have a height of 74–78 cm for men and 70–74 for women assuming that the chairs are fully adjustable and foot rests are available.* These figures are slightly higher than most standard specifications recommending desk top heights of between 72 and 75 cm, which are certainly not ideal for tall male employees.

*Vertical and horizontal leg room*

It is important that office desks allow plenty of space for leg movement and it is an advantage if the legs can be crossed without difficulty. For this reason there should be no drawers above the knees, and no thick edge to the desk top. The table top should not be thicker than 3 cm and the space for legs and feet under the table should be at least 68 cm wide and 69 cm high.

Many employees, especially those who lean back, periodically like to stretch their legs under the table. It is therefore necessary to leave enough depth as well. *At knee level the distance from table edge to back wall should not be less than 60 cm increasing to 80 cm at ground level.* These recommendations are also valid for workplaces with typewriters or VDTs.

*Typing desks calls for height adjustability*

The above mentioned classical guideline, requiring a straight upright trunk position with the elbows at the sides and bent at right angles has for a long time been the basis for the design of typing desks. Since the height of the keyboard defines the working level, the middle row (or so-called home row) should be at about elbow level. However, the need for such a low desk conflicts with the necessity for enough knee space under the table. This can be a limiting factor and *calls for height-*

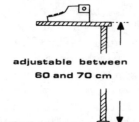
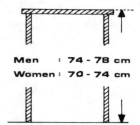

**Figure 50 Recommended desk-top heights for traditional office jobs.**
*Left: Range of adjustability for typing desks.*
*Right: Desk-top heights for reading and writing without typewriter.*

*adjustable typing desks.* Indeed, *today most experts recommend the height of such desks to be adjustable between 60 and 70 cm.* These recommendations are illustrated in Figure 50.

The height of a non-adjustable typing desk is a very problematical dimension. *Until now most experts have recommended a fixed top height of 65 cm.* This recommendation is based on two assumptions which are questionable for the following reasons:

1. The assumed straight upright trunk posture is not normally adopted by typists if the work lasts for a few hours or more. They often lean back or forwards in order to relax the back muscles or to get a suitable viewing distance.
2. The old mechanical typewriters required stroke forces of several hundred grams. Modern electrical typewriters need much lower stroke forces of 40–80 g; the keys are easily operated by the fingers alone and no dynamic muscular effort is required from the forearms and hands, which therefore need not be kept in a horizontal plane!

*Thus it is concluded that the recommended height of 65 cm for a fixed typing desk has yet to be demonstrated in a convincing manner by taking into account the preferred and actually assumed postures of typists and by providing operators with comfortable office chairs as well as with electric typewriters.*

## 7.3. Field studies on musculoskeletal troubles of office employees

In the last decade many studies have shown that constrained postures and musculoskeletal discomfort are often observed in office jobs involving regular work with machines, such as full-time typing, operating accounting or punching machines as well as working with VDTs. These publications are listed in the bibliography (5, 30, 35, 41, 50, 51, 84, 93, 104, 107, 121, 124, 125, 128, 137, 151, 160, 177, 186, 192, 193).

Reviews of some of the field studies on VDT workstations were published by Dainoff (42), the US National Academy of Sciences (153), Helander et al. (86) and Läubli (123).

*Frequent symptoms: discomfort in neck and shoulders*

All of the above mentioned authors *reported complaints* about physical discomfort, often localized in the neck–shoulder–arm area. However, in some of the studies (104, 192, 193), no significant difference was ascertained between the VDT and the chosen control groups, which accounts for the fact that the respective conclusions differ from those of other authors.

The studies on musculoskeletal discomfort of VDT operators have the same weaknesses as those discussed in

Section 6.1: the control groups are in most cases not comparable. In fact, a VDT job cannot be compared to a 'non-VDT job' without reservation; there are always differences apart from the use or non-use of a VDT. Nevertheless it is worth while analysing some of the studies more closely.

*Field studies on telephone operators*

Starr et al. (192), in the telephone operator survey already mentioned in Section 6.1, also studied the musculoskeletal complaints of 145 directory enquiry operators. Of 15 reported items only one, namely 'neck discomfort', was statistically significant: 65% of the VDT operators reported discomfort versus 48% of the control group. The complaints about discomfort in shoulders and back were also higher in the VDT group but statistically not significant. An index of the total number of physical complaints out of 15 was calculated. The mean value of this index was considerably higher for the VDT group.

*Hardly any difference to control groups...*

However, surveying the age differences the authors were not able to confirm the statistical significance between the compared groups. They concluded that "replacing paper documents with VDTs need not adversely affect comfort and morale of office workers".

In a more recent study, Starr (193) repeated a similar field study on 211 VDT users and 148 control subjects who handled requests from residential telephone customers. The groups exhibited few differences in the likelihood and intensity of on-the-job physical discomfort. The incidence of troubles in the neck and in the shoulders were of the same order of magnitude in both groups.

*...but advanced furniture reduced physical discomfort*

In a follow-up investigation Shute and Starr (182) studied the effect of advanced furniture designed for VDT workstations, again on directory enquiry operators. Details of this study will be discussed later, but the main result shall be mentioned here: the use of an advanced table and chair was associated with a significant improvement of comfort, the incidence of reported discomfort being considerably reduced in the neck, shoulders, upper arms, wrists and other parts of the body. This study shows that the reported physical discomforts of VDT operators are, at least to some extent, a reality, since the symptoms vanished partially with the use of advanced furniture reducing constrained postures.

*Four studies on VDT and control groups*

The results of four field studies conducted by Hünting et al. (93), Maeda et al. (137), van der Heiden et al. (84) and Läubli et al. (124) reveal the crucial importance of control groups. These studies on musculoskeletal discomfort could be compared to each other since the same questionnaire was used in

**Table 21** Some characteristics of the eight groups involved in four studies.

| Group | VDT | Survey | n | Age ($\bar{x} \pm$ S.D.) | Women (%) | 6h/day at keyboard or terminal (%) |
|---|---|---|---|---|---|---|
| Data entry tasks | + | (93) | 53 | $30 \pm 8$ | 94 | 81 |
| Conversational tasks | + | (93) | 109 | $34 \pm 12$ | 50 | 73 |
| CAD operators | + | (84) | 69 | $33 \pm 8$ | 23 | 20 |
| Mechanical and wiring-board design | – | (84) | 19 | $34 \pm 12$ | 31 | – |
| Flight reservation in an airline | + | (124) | 45 | $29 \pm 7$ | 58 | 100 |
| Accounting operators | – | (137) | 119 | $21 \pm 3$ | 100 | 80 |
| Full-time typists | – | (93) | 78 | $43 \pm 13$ | 95 | 65 |
| Traditional clerical work | – | (93) | 55 | $28 \pm 11$ | 60 | 30 |

(93) = Hünting *et al.* (137) = Maeda *et al.* (84) = van der Heiden *et al.* (124) = Läubli *et al.*; VDT+ = working with VDT; S.D. = standard deviation.

each of the four surveys. Some characteristics of the eight groups of employees involved are presented in Table 21.

The jobs of the eight groups can be described as follows:

1. *Data entry task.* Full-time numeric data entry with the right hand; 12 000 – 17 000 strokes/h. Vision is mainly directed at the documents.
2. *Conversational task.* Payment transactions in two banks. Both hands operate the keyboard. Vision is directed about 50% of the time at the screen and 50% at the source documents. Low stroke speed.
3. *CAD operating.* Mechanical design, printed circuit-board and electrical schematics design. Tablets are used for about 40% and the keyboard for about 20% of the working time. Vision is directed for about 60% of the time on the screen. Great diversity of body movements.
4. *Mechanical and circuit-board design.* The same job as CAD operating, but without a VDT.
5. *Flight reservation in an airline.* Conversational type of job. No source documents; all information is given on the screen. 70% of the machines have a fixed keyboard.
6. *Accounting machine operating.* Full-time numerical data entry, taken from coupons and typed with the right hand only. Vision is mainly on the coupons. 8 000 – 12 000 strokes/h.
7. *Full-time typing.* Partly copying documents, partly using dictating machine. Vision mainly on documents. High typing speed.
8. *Traditional clerical work.* Payment transactions in a branch office of a bank (without VDT); keyboards are used only occasionally; great diversity of body movements.

Some of the results of the questionnaires are reported in Table 22.

Table 22  Incidence of "almost daily pains" in the neck—shoulder—arm and hand areas, reported from four field studies.

|  |  |  | Subjects with "almost daily" pains in | | | |
|---|---|---|---|---|---|---|
| Group | VDT | Survey | Neck (%) | Shoulder (%) | Right arm (%) | Right hand (%) |
| Data entry tasks | + | (93) | 11 | 15 | 15 | 6 |
| Conversational tasks | + | (93) | 4 | 5 | 7 | 11 |
| CAD operators | + | (84) | 3 | 3 | 3 | 3 |
| Mechanical and wiring-board design | — | (84) | 2 | 1 | 2 | 0 |
| Flight reservation in an airline | + | (124) | 7 | 11 | 4 | 6 |
| Accounting operators | — | (137) | 3 | 4 | 8 | 8 |
| Full-time typists | — | (93) | 5 | 5 | 4 | 5 |
| Traditional clerical work | — | (93) | 1 | 1 | 1 | 0 |

r = Right; VDT+ = working with VDTs. The number of subjects in each group = 100%. Survey references as Table 21.

*Physical discomfort in the neck, shoulder and arms*

Because of the heterogeneous composition of the eight groups a statistical analysis of the results would not make sense. Thus they will only be looked at with 'common sense'. The figures reveal that musculoskeletal troubles in the neck—shoulder—arm—hand region were observed in each group. The highest figures, though, were found in the group operating data entry terminals, whereas the lowest figures were reported by CAD operators and their control group (draughtsmen for mechanical and other design) as well as by the group engaged in traditional clerical work.

It is striking that the three groups whose work is characterized by a great diversity of body movements clearly present the lowest incidence of physical discomfort. On the other hand, it seems that jobs involving repetitive work on machines are likely to cause physical discomfort in the neck—shoulder—arm—hand region. Some of the VDT jobs certainly belong to this type of work, but not all of them. Therefore, *the determining causal factor is not the question of 'use' or 'non-use' of VDTs.*

*Discomfort and key strokes*

An interesting contribution to this discussion is a study by Läubli (123), who compared the incidence of pains in the neck and arms of eight groups using keyboards. Seven of these groups were identical with those listed in Tables 21 and 22. Furthermore, the approximate typing speed was assessed for each of the eight groups. The results are given in Figure 51, and show that the difference between use and non-use of VDTs among the eight groups is much less important than the differences concerning the daily key strokes.

# Ergonomic design of VDT workstations

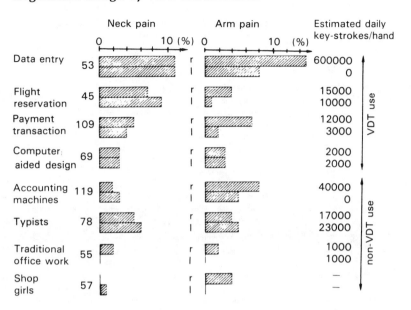

Figure 51 Incidence of "almost daily pains" in the neck and arm in eight different occupational groups, related to the estimated number of daily key-strokes.
*Abscissa: percentage of subjects of each group. Differences of more than 5% are significant. According to Läubli (123).*

These results confirm what was said above: *a VDT as such does not cause physical discomfort, it is the way it is used that is responsible for constrained postures and the ensuing troubles.*

*Medical findings* — Questionnaires reflect subjective feelings but they give no information on the medical nature of the troubles. That is why Läubli applied investigation methods which are used in rheumatology (125) and include the assessment of joint and back mobility, painful pressure points at tendons or other characteristic locations and painful reactions to muscle palpation. He noticed a high correlation between the medical findings of two examining doctors and between these medical findings and self-rated physical discomfort (125). The study includes all subjects of the field study (93). Some of the results are presented in Table 23 and in Figure 52.

The results of Table 23 confirm the complaints reported in Table 22: medical findings, indicating musculoskeletal troubles in muscles, tendons and joints are frequent in the groups using data entry terminals and among full-time typists, whereas the group with traditional office work consisting of many different activities and movements shows the lowest figures. The palpation findings in shoulders, listed in Figure 52, disclose a similar distribution of symptoms.

Table 23 Incidence of medical findings in the neck–shoulder–arm area of office employees.

| Medical findings | Data entry tasks ($n=53$) (%) | Conversational tasks ($n=109$) (%) | Full-time typists ($n=78$) (%) | Traditional office jobs ($n=54$) (%) |
|---|---|---|---|---|
| Tendomyotic pressure pains in shoulders and neck | 38 | 28 | 35 | 11 |
| Painfully limited head movability | 30 | 26 | 37 | 10 |
| Pains in isometric contractions of forearm | 32 | 15 | 23 | 6 |

$n$ = Number of subjects, equal to 100% in each group. According to Hünting et al. (93).

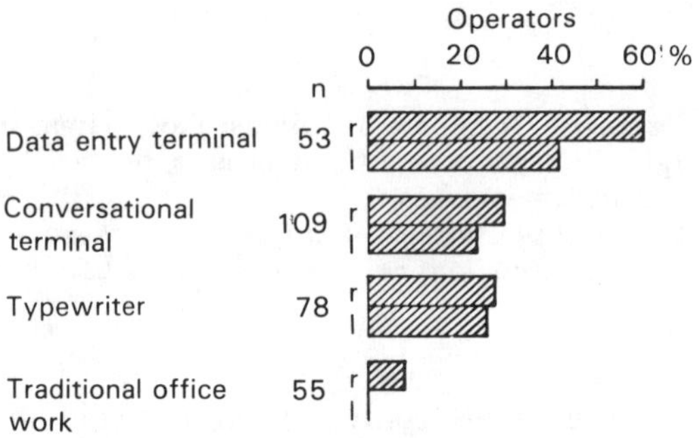

Figure 52 Palpation findings in the shoulders of four groups of office workers. Clinical symptoms are very frequent at data entry terminals and rare in traditional office work.
*Painful pressure points are at the tendons, joints and muscles. r = right; l = left; n = number of examined operators. Differences between groups was significant at p <0.01; Kruskal-Wallis test. According to Läubli (121).*

The complaints as well as the medical findings must be taken seriously, especially since 13–27% of the examined employees had consulted a doctor for this reason.

## 7.4. Postures, workstation characteristics and physical discomfort

*A study on accounting machine operators*

A study on accounting machine operators (92, 137) revealed interesting relationships between certain constrained postures and the incidence of musculoskeletal discomfort. Fifty-seven female accounting operators were selected for the

*Ergonomic design of VDT workstations*

**Figure 53  A view of an accounting machine operator with a typical working posture.**
*The neck is bent, the right arm is elevated with a lateral abduction of the hand and the left elbow is supported by the desk. According to Hünting et al. (92) and Maeda et al. (137).*

postural study. They were set the task of reading figures from coupons and of typing them into a numeric keyboard. With the left hand the operator had to turn the coupons over while resting the left elbow on the desk. The face was directed towards the coupons causing some forward bending of the neck. The right hand was exclusively used to operate the numerical keyboard. The job is very similar to a data entry task of VDT operators. Figure 53 shows the typical working posture of an accounting machine operator.

The number of daily working hours was 8.5, the work on accounting machines lasting 5–6 hours. The keying speed varied between 8000 and 12 000 strokes/h. A questionnaire illustrated with anatomical drawings was used to rate the various symptoms of physical discomfort. The assessment of postures covered the visual distance and several angles characterizing the positions of trunk, head, arms and hands. The postures of the operators were very constant; the repetition of measurements gave no significant variance.

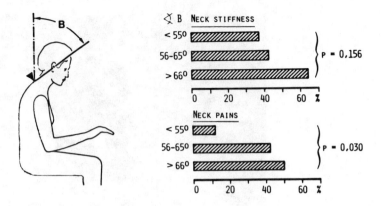

**Figure 54** Incidence of physical discomfort in the neck related to the ranges of neck–head angles (B) of 57 accounting machine operators.
*Statistical analysis by the Mann–Whitney U Test.*

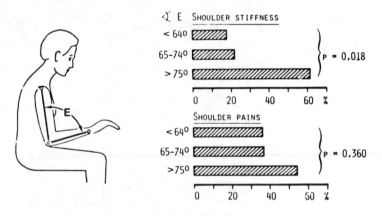

**Figure 55** Incidence of physical discomfort in the right shoulder related to the ranges of the right elbow angles (E) of 57 accounting machine operators.
*Statistical analysis by the Mann–Whitney U Test.*

*Relationships between postures and physical discomfort*

To check possible effects of posture on the incidence of reported discomfort, the 57 subjects were divided into three groups of different postural characteristics. The comparison of these data showed interesting relationships between postural angles and incidence of discomfort. A selection of results is offered in Figures 54–56.

Although not all of the reported relationships are statisti-

# Ergonomic design of VDT workstations

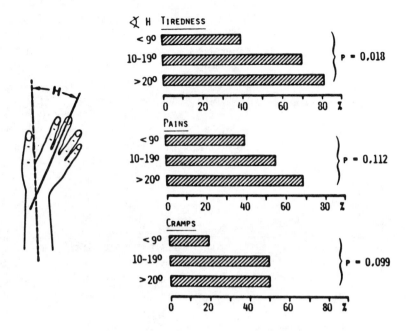

Figure 56 Incidence of physical discomfort in the right hand related to the ranges of lateral (ulnar) abduction of the right hand (H) of 57 accounting machine operators.
*Statistical analysis by the Mann–Whitney U test.*

cally significant, the general tendency is obvious and leads to the following conclusions:

1. Stiffness and pains in the neck increase with an increasing degree of forward bending of the head.
2. Stiffness and pains in the right shoulder increase with increased opening of the elbow and lowering of the right hand. It should be remembered that the operators could not rest wrists or hands on a proper support.
3. Tiredness, pains and cramps in the right hand increase with an increasing degree of lateral (ulnar) abduction of that hand. See the right hand position in Figure 53.

In the above mentioned field study on VDT operators (93, 121, 123) several significant relationships were discovered between the design of workstations or postures on the one hand and the incidence of complaints or medical findings on the other. These results can be summarized as follows: physical discomfort and/or the number of medical findings in the neck–shoulder–arm–hand region are likely to increase when:

1. The keyboard level above the floor is too low.
2. Forearms and wrists cannot rest on an adequate support.

3. The keyboard level above the desk is too high.
4. Operators have a marked head inclination.
5. Operators adopt a slanting position of the thighs under the table due to insufficient space for the legs. This is illustrated in Figure 57.
6. Operators disclose a marked lateral (ulnar) abduction of the hands when operating the keyboard; this is shown in Figure 58, where the subjects are divided into two subgroups: one with an angle of ulnar abduction of the hand of more than 20°, the other with an abduction of less than 20°.

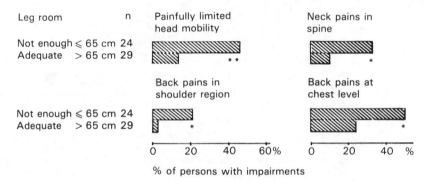

Figure 57 **Vertical leg room and physical discomfort in 53 VDT operators at conversational terminals.**
Leg room = distance from the floor to the lower edge of the desk.
n = Number of operators; *$p < 0.05$; **$p < 0.01$; Mann-Whitney U test.
According to Läubli and Grandjean (123).

*Workstation and spinal disc load*

Another study, carried out by Cantoni et al. (30), deals with effects of workstation design on the back load. This investigation was conducted on 300 VDT operators (50% males) working in the switchboard control room of the Italian National Telephone Company in Milan where a gradual transformation of workstations from the traditional electromechanical switchboard to a VDT-operated switchboard was in progress. The old switchboard had a fixed working surface and a vertical panel in which the operators inserted the jacks; the chairs had fixed backrests and leg room was too small. The new workstation has a screen which is adjustable in a vertical, horizontal and sagittal plane; the keyboard is recessed in the table; a large space of 25 cm is left to rest the wrists and forearms; and the chairs are provided with a flexible backrest 28 cm high. The survey revealed a high incidence of troubles in the cervical spine area (ca. 55% of them over 36 years old) and in the lumbar region (ca. 55% of them older than 36) in operators working at the electromechanical switchboard. In order to study the influence of the new workstation on

Ergonomic design of VDT workstations

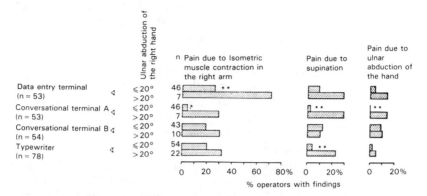

**Figure 58** Incidence of medical findings in the right forearm of two groups of subjects presenting different degrees of lateral (ulnar) abduction of the right hand.
*Terminal A: large area of rest for wrists and hands, movable keyboard. Terminal B: narrow rim of about 3 cm to rest the balls of the thumbs, keyboard recessed in the desk surface. p = significance of differences between the angle groups:* *=p < 0.05; **=p < 0.01; *Mann-Whitney U test.*

postural behaviour a limited group of four subjects, two males and 2 females aged between 37 and 44 years with a body height between 159 and 170 cm, were selected. With a work-sampling procedure the characteristic trunk positions were rated and the duration of each posture determined. For each postural variant the load on the intervertebral disc L3 was

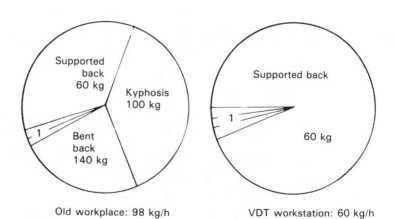

**Figure 59** Average working time with the corresponding back postures of four subjects on an electromechanical telephone switchboard (on the left) and on a VDT operated switchboard (on the right).
*The figures inside the circle indicate the calculated load on the intervertebral disc (L3). The figures below the circles indicate the mean intervertebral load per hour. 1 = various other trunk postures. According to Cantoni et al. (30).*

calculated according to a procedure by Molteni et al. (143) and an unpublished modification of it for supported upper limbs by Occhipinti et al. (163). In Figure 59 the observed duration of each posture and the mean load on L3 is shown for the old and the new workstation.

Two results should be pointed out here: the alteration of the workstation and the change of posture caused a decrease in the calculated average load on the third lumbar disc (L3) from 97.5 kg/h to 60 kg/h. This is a significant improvement and a substantial relief to the lumbar region. Another important result is the increase in prolonged periods of the back leaning against the backrest (from about 30% to nearly 100% of the time). The 'bent back' and 'kyphosis' positions nearly disappeared. On the other hand, a marked decrease of diversity of postures was observed at the VDT workstation, an effect considered a disadvantage by the authors. For this reason they recommend a 15-minute break every two hours.

*Trunk position and disc loads*

The same group of authors (163) recently analysed the effects of resting the upper limbs on the desk. They observed that the supporting forces varied, depending on upper body posture, between 3.5% and nearly 15% of the body weight. Supporting the upper limbs significantly reduced the load on the intervertebral discs. The disc loads of various trunk postures with unsupported and supported upper limbs are presented in Figure 60.

The results show the importance of a proper support for the upper limbs of office employees using keyboards.

| Unsupp. | 100 kg | 120 kg | 125 kg | 131 kg | 189 kg | 210 kg |
| Supp.   | 91 kg  | 88 kg  | 10 kg  | 97 kg  | 137 kg | 117 kg |

Figure 60 Discal loads of an average subject (weight 70 kg and height 170 cm) with different trunk postures with unsupported and supported upper limbs.
*According to Occhipinti et al. (163).*

*Inclined desk-top and trunk posture*

In this context one should also take notice of a study by Bendix and Hagberg (14) who examined the effects of inclined desk-tops on 10 reading subjects. With increasing desk inclination the cervical as well as the lumbar spine were extended and the head and trunk assumed a more upright posture. The electromyograph of the trapezius muscle (shoulder-lifting muscle) disclosed no change in muscle strain.

A rating of acceptability for both reading and writing on either desk inclination favoured a steep slope of the desk for reading whereas the opposite was preferred for writing. The authors conclude that the reading material should be placed on a sloping desk and the paper used for writing on the horizontal table-top between the desk and the person. A separate sloping desk placed on a level table should be preferred to inclining the total table-top since a slope of more than 10° usually causes paper and pencils to slide off.

## 7.5. Orthopaedic aspects of the sitting posture

*A historical perspective*

Many primitive peoples had no knowledge of seats of any kind; they crouched, knelt or squatted on the ground. Seats originated, at least in part, as status symbols; only the chief had the right to be raised by a seat. Hence the gradual development of ceremonial stools, which indicated status by their size and decoration. Figure 61 shows such an example. The climax of this development in cultural history was reached with the splendour of a throne. This status function has persisted to the present day. Anyone doubting this should have a look at a prospectus of a company selling office furniture and notice that there is a type of chair corresponding to each salary level! Thus there is a wooden chair for typists; for the senior clerk one that is thinly upholstered; a thickly upholstered chair for the office manager and for directors a swivel armchair upholstered in leather.

*Chairs were first of all status symbols*

*Physiological reflections emerged*

At the beginning of the present century it was gradually realized that well-being and efficiency could be improved and fatigue reduced if people could sit at their work. The reason is a physiological one: as long as a person is standing a certain static muscular effort is required to keep the joints of the feet, knees and hips in a fixed position; such muscular effort ceases when the person sits down.

This led to an increased application of medical and ergonomic ideas to the design of seats for work. The developments gained in importance as more and more people sat down at their work; today about three-quarters of all workers in industrial countries have sedentary jobs.

*Pros and cons of sitting*

The advantages of a sedentary posture at work are:

Taking the weight off the legs.
Possibility of avoiding unnatural postures.
Reduced energy expenditure.
Lower demands on blood circulation.

These advantages must be set against certain drawbacks: prolonged sitting leads to a slackening of the abdominal muscles

Figure 61 The seat of the mayor of Bern (Switzerland) created by the cabinet-maker M. Funk in 1735.
*This beautiful chair was certainly, above all, a status symbol for the distinguished major who was allowed to use it. (Historisches Museum Bern, Switzerland.)*

('sedentary tummy'), and to a curvature of the spine which, in turn, is unfavourable for the organs of digestion and breathing.

But the most serious problem involves the spine and the muscles of the back, which in many sitting positions are not only not relaxed but positively stressed in various ways.

*Intervertebral disc*

About 60% of adults have backaches at least once in their life and the most common cause of this is disc injuries. An intervertebral disc is a sort of cushion which separates two

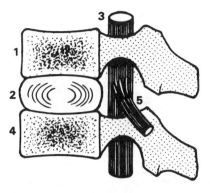

Figure 62 Diagram of a section of the spine. The disc (2) lies between two vertebrae, (1) and (4); behind, the spinal cord (3) and a nervous tract (5). The disc is like a cushion which gives flexibility to the spine.

vertebrae and collectively they give flexibility to the spine. A disc consists of a viscous fluid enclosed by a tough, fibrous ring, which encircles the disc. A schematic representation of a disc between two vertebrae and its connections with the spinal cord and the nervous tracts is given in Figure 62.

*Disc injuries*

For reasons that are still unknown, intervertebral discs may degenerate and lose their strength: they become flattened and in advanced cases the viscous fluid may even be squeezed out. This impairs the mechanics of the vertebral column and allows tissues and nerves to be strained and pinched, leading to various diseases of the back, to lumbago, to sciatic troubles and even to paralysis of the legs.

Unnatural postures, weight-lifting and bad seating can speed up the deterioration of the discs, resulting in all the ailments mentioned above. For this reason many orthopaedists have started to concern themselves with the medical aspects of the sitting posture, including Akerblom (2), Schoberth (179), Yamaguchi (211), Keegan (100), Nachemson (149), Andersson (4) and Krämer (110).

*Orthopaedic research*

A very important contribution was made by the Swedish orthopaedists Nachemson (149) and Andersson (4), who developed sophisticated methods to measure the pressure inside a disc during a variety of standing and sitting postures. They emphasize that increased disc pressure means that the discs are being overloaded and will wear out more quickly. Therefore disc pressure is a criterion for evaluating the risk of disc injuries and backaches.

*Disc pressure for four postures*

The effects of four different postures on nine healthy subjects are shown in Figure 63. The results disclose that *the disc*

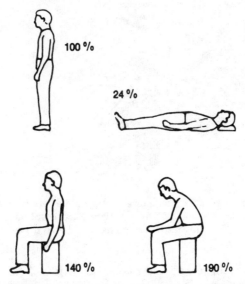

Figure 63 **The effect of four postures on the intervertebral disc pressure between the 3rd and 4th lumbar vertebrae.**
*The pressure measured when standing is taken as 100%. According to Nachemson and Elfström (149).*

*pressure is greater when sitting than when standing.* The explanation lies in the mechanism of the pelvis and sacrum during the transition from standing to sitting:

The thigh acts as a lever.
The upper edge of the pelvis is rotated backwards.
The sacrum turns upright.
The vertebral column changes from a lordosis to either a straight or a kyphotic shape.

*The spine when standing and sitting*

It should be recalled here that lordosis means that the spine is *curved forward, as it normally is in the lumbar region when standing erect;* and kyphosis describes the backward curving, which is normal in the thoracic region when standing upright. These effects of standing and sitting posture are illustrated in Figure 64.

*The backward position of the pelvis puts the spine into a state of kyphosis which, in turn, increases the pressure within the discs.*

*Erect or relaxed sitting posture*

This is the reason why many orthopaedists still recommend an upright sitting posture, as this holds the spine in the shape of an elongated S with a lordosis of the lumbar spine. In fact, disc pressure is lower in such a posture than when the body is curved forward with a predominant kyphosis in the lumbar and thoracic regions. One of the recent advocates for an upright trunk at working desks is Mandal (139), who recommends higher seats and higher sloping desks which

*Ergonomic design of VDT workstations*

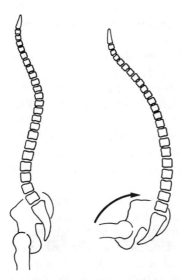

**Figure 64 Rotation of the pelvis when changing from a standing to a sitting posture.**
*Left: Standing upright.*
*Right: Sitting down. Sitting down involves a backward rotation of the pelvis (indicated by the arrow), bringing the sacrum to an upright position and turning the lumbar lordosis into a kyphosis.*

automatically leads to a more upright posture with reduced forward bending of the back.

In the same vein are the 'Balans' seats from Norway, which induce a half sitting half kneeling posture. The seat surface is titled forward (24°), and a support for the knees prevent a forward sliding of the buttocks. The result is a wide opening of the hip angle (between legs and trunk) and a pronounced lumbar lordosis with a straight upright trunk posture. However, Krueger (117) tested four models and found that the load on knees and lower legs is too high and sitting becomes painful

**Figure 65 Electrical activity in the back muscles when sitting upright, and in a relaxed posture, slightly bent forward.**
*Sitting upright involves a considerable electrical activity, revealing the static effort imposed upon the muscles of the back. According to Lundervold (135).*

after a while. (Some subjects even refused to sit longer than 2 hours.) With desks of 72–78 cm height the effect of a lumbar lordosis is reduced since the subjects cannot avoid bending the trunk forward. Drury and Francher (47) tested a similar forward-tilting chair which elicited mixed responses, with complaints of leg discomfort from VDT users. Overall, the chair was no better than conventional chairs and could be worse than well-designed office seats. Looking at these forward-tilting chairs, the question arises whether they could not be of greater use as physiotherapeutical exercise than in an office!

The orthopaedic advice of an upright trunk posture conflicts with the fact that a slightly forward or reclined sitting posture relieves the strain on the back muscles and makes sitting more comfortable. This was confirmed, in part, by electromyographic studies; the result of a classical experiment carried out by Lundervold (135) in 1956 is shown in Figure 65.

*Slightly forward bent trunk holds body weight in balance*

From this it emerges that a relaxed posture, with a slightly forward bent trunk, holds the weight of the body in balance with itself. This is the posture that many people adopt when they make notes or read in a sitting position, because it is relaxing and exerts a minimum of strain on the muscles of the back. The visual distance for good readability might in some cases be another reason for a slightly forward bent trunk. Thus there is a 'conflict of interests' between the demands of the muscles and those of the intervertebral discs: while the discs prefer an erect posture, the muscles prefer a slight forward bending.

*Nutrition of intervertebral discs*

Here the interesting investigations of Krämer (110) must be noted, offering a detailed study of the nutritional needs of intervertebral discs. The interior of a disc has no blood supply and must be fed by diffusion through the fibrous outer ring. Krämer has produced evidence that pressure on the disc creates a diffusion gradient from interior to exterior, so that tissue fluid leaks out. When the disc pressure is reduced, this gradient is reversed and tissue fluid diffuses back, taking nutrients with it. This seems to prove that in order to keep the discs well nourished and in good condition they need to be subjected to frequent changes of pressure, a kind of pump mechanism. From a medical point of view, therefore, *an occasional change of posture from bent to erect or from leaning back to an upright position and vice versa must be beneficial.*

*Increased seat angle reduces disc load*

Andersson and Ortengren (4), mentioned above, also studied the effects of seat angle and different postures at working desks on disc pressure. Simultaneously the electrical activity of back muscles was recorded to measure the static load. The effects of different postures are described in Figure 66. The

*Ergonomic design of VDT workstations* 125

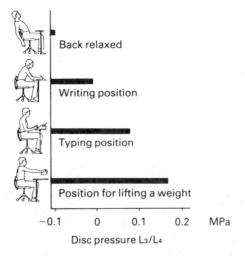

**Figure 66 Effects of various sitting postures on disc pressure.**
*(1 MPa = 10.2 kp/cm². ) $L_3$ and $L_4$ = third and fourth lumbar vertebrae. Zero (0) on the pressure scale is a relative reference value for a seat angle of 90°. Absolute values at the reference level zero were about 0.5 MPa (= 5 kp/cm²). According to Andersson and Ortengren (4).*

results reveal that leaning back as well as bending forward with supported upper limbs (writing posture) are favourable conditions for the disc pressure.

The effects of the seat angle are represented in Figure 67, and results are clear: by increasing the seat angle both disc pressure and muscle strain are reduced.

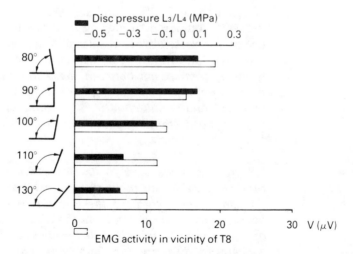

**Figure 67 Effect of seat angle (i.e., between seat and backrest) on disc pressure and on the electrical activity in the back muscles recorded at the level of the eighth thoracic vertebra (T8).**
*For more details see Figure 66. According to Andersson and Ortengren (4).*

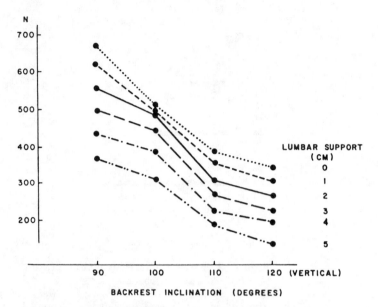

**Figure 68 Effects of different sizes of lumbar support and of increasing seat angles on disc pressure.**
*The size of the lumbar support is defined as the distance between the front of the lumbar pad and the plane of the backrest. The backrest inclination is defined as the angle between the seat and the backrest.*

*A proper lumbar pad relieves disc strain*

Another study by the same authors (4) showed that a proper lumbar support also resulted in a decrease of disc pressure. These results are reported in Figure 68. Further studies on the adjustment of back support of office chairs at different lumbar levels showed that fixing the back support at the level of the fourth and fifth lumbar vertebrae slightly decreased pressure compared to placing it at the first and second. The use of arm rests always results in a decrease of disc pressure, less pronounced, however, when the backrest—seat angle is large. *A comparison between these findings shows that the disc load of a person leaning back at an angle of between 110 and 120° and supplied with a 5 cm lumbar pad is even lower than that of a standing posture with the advocated lordosis of the lumbar region.*

*Conclusion from orthopaedic research*

All these studies lead to an important conclusion: *resting the back against an inclined backrest transfers a significant portion of the weight of the upper part of the body to the backrest and reduces strain on the discs and muscles. In view of the design of chairs it is deduced that optimum conditions concerning disc pressure and muscular activity are given when the backrest has an inclination of 110 or 120° and a 5 cm thick lumbar pad.*

*The cervical spine*

Another part of the spine is as important as the lumbar region: the cervical spine, or spine of the neck, consisting of the first seven vertebrae. Like the lumbar spine, this is a very mobile segment, also showing a lordosis when standing upright. The cervical spine is a delicate part and prone to degenerative processes and arthrosis. A great number of adults have neck troubles due to injuries of the cervical vertebrae and discs, generally referred to as cervical syndrome. The most common symptoms are painful indurations or cramps of the shoulder muscles, pains and reduced mobility in the cervical spine and sometimes painful radiation in the arms, ailments which are also called cervicobrachial syndrome. In Japan these cervicobrachial disorders are considered an occupational disease since they are often observed among key-punchers, assembly plant workers, typists, cash register operators and telephone operators (138). Recently several authors have discovered physical troubles among VDT operators which fit the above description of cervicobrachial disorders (121, 160). As shown in Figure 54 physical discomfort in the neck increases with an increasing degree of forward bending of the head.

*Posture of the neck and head*

The posture of the neck and head is not easy to assess since seven joints determine the great mobility of this part of the body. In fact, it is possible to combine an erect or even lordotic neck with a downward bent head or a forward flexed neck with an upwards directed head. Some authors define the neck–head posture by measuring an angle between a line along the neck related either to a horizontal (or vertical) or to a line along the trunk. Most authors consider such a neck–head angle of 15° to be acceptable. Chaffin (33) determined the neck–head angle to the horizontal and observed on five subjects that the average time to reach a marked muscle fatigue was shorter with increasing neck–head angles. The author concludes that localized muscle fatigue in the neck area can be a preliminary sign of other more serious and chronic musculoskeletal disorders and that the inclination angle of the head should not exceed 20–30° for any prolonged period of time. Let us note here that neck pains can vanish overnight when source documents are presented on raised reading standings instead of laying them flat on the desk top.

Another way of assessing the posture of the neck and head is by measuring the angle between a line running from the seventh cervical vertebra to the earhole and a vertical line. Figure 54 (p. 114) illustrates this type of neck–head angle.

*'Normal' line of sight...*

A further approach to the problem of neck–head postures is the assessment of the 'normal' line of sight. The direction of sight is determined firstly by the movement of the eyeballs and secondly by the posture of the neck and head. Eye movement within 15° above and below the normal line of sight is still comfortable. This means that regular viewing tasks

*... is 10–15° below the horizontal plane*

should be within a 30° cone around this principal line of sight. If a target lies outside this cone it is assumed that the neck–head mechanism is involved.

What is the 'normal' line of sight? Most authors today agree that it is 10°–15° below the horizontal plane. Although not stated explicitly most authors consider the 'normal' line of sight a resting condition of the eye. *For this reason a display or any other visual target should be placed within a viewing angle of 5° above and 30° below the horizontal plane.* This plane may refer to the eye rotation as such or merely to the slight inclination of the head. The 'normal' line of sight with the cone of easy eye rotation is shown in Figure 69.

Two recent studies confirmed that the 'normal' line of sight is also the preferred line of sight of operators watching a screen. VDT operators who could fix the height of the screen as desired came up with an average figure for the preferred viewing angle of 9° below the horizontal (73). These results are reported in Table 25 and will be discussed in detail in Section 7.7.

In a recent study, carried out by Bhatnager *et al.* (17), performance, discomfort and posture were analysed in relation to different height positions of a screen. Twelve subjects inspected printed circuit boards for 3 hours and had to detect breaks in the circuit. Three screen heights—61 cm, 153 cm and 244 cm — determined corresponding postures and lines of sight. The height of 153 cm was some 20 to 30 cm above eye level and therefore closest to the normal line of sight.

As working time went on, subjects began to lean forward, changed the position more often, and commented on increasing discomfort, and the quality of their performance deteriorated. The screen height had an effect both on frequency of posture changes (a sensitive indicator of postural

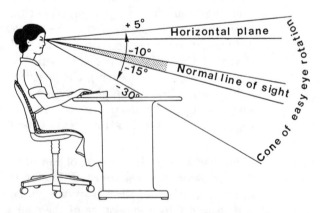

**Figure 69** The 'normal' line of sight with the range of easy eye rotation.

fatigue) and on degree of discomfort. The least comfortable position and the worst performance were effected by the high screen position. This was farthest away from a 10° downward line of sight. Performance and physical comfort were best for the middle screen height closest to the 'normal' line of sight.

*Differing opinions*  A 'normal' line of sight of 10–15° below the horizontal contradicts an earlier recommendation of Lehmann and Stier (130) who found that seated subjects preferred an average viewing angle of 38° below the horizontal, whereby approximately half of this angle was attributed to the downward bent head. This controversial result is most probably due to special experimental conditions including short test durations.

*Line of sight related to head position*  In a recent study Hill and Kroemer (87) investigated the preferred line of sight by measuring its angle to the so-called Frankfurt plane, which defines the spatial orientation of the head by anthropometric landmarks. The Frankfurt plane is established by a line passing through the right earhole and the lowest edge of the right orbit. Naturally this plane moves only with the head. It coincides with the horizontal when the head is held straight and erect. Thirty-two subjects were tested in several experiments, the total time of which lasted about two hours. The seat was provided with a high backrest and could be inclined backwards from 90° to 105 and 130°. The subject's head was placed against the headrest and was not flexed. The results showed an overall mean angle of 34° below the Frankfurt plane with a large range between 14° above and 71° below the reference plane. When the subjects leaned back the range of the Frankfurt plane was raised to between 15° above and 40° below.

Relating the line of sight to the horizontal (and not to the Frankfurt plane) yields the following results:

| Backrest angle | Line of sight related to horizontal |
|---|---|
| 90° (upright posture) | 28.6° below |
| 105° (leaning back) | 19.4° below |
| 130° (leaning back) | 8° above |

These results are a little confusing; they are not in agreement with the general opinion of a 10° or 15° declination of the line of sight from the horizontal. Again it must be assumed that the special experimental conditions and the relatively short duration of each test might to some extent be responsible for these deviating results.

*Conclusions for posture of neck and head*  The present state of knowledge suggests that the head and neck should not be bent forward by more than 15°, otherwise fatigue and troubles are likely to occur. The preferred line of *sight lies on average between 10 and 15° below the horizontal plane and this corresponds very well to the preferred viewing angles of VDT operators watching their screen.*

## 7.6. Ergonomic design of office chairs

*Preferred seat profiles*

Ergonomic research in the field of sitting posture has either been related to preferred seat designs or to the sitting behaviour of users. Experiments with a 'sitting machine' (68) and with a great variety of moulded seat shells of different profiles (69) led to the shape of a multipurpose chair shown in Figure 70. This preferred seat shape is characterized by a slightly moulded seat surface in order to prevent the buttocks

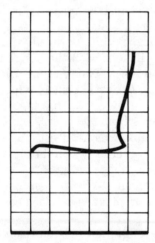

**Figure 70 A seat profile of a multipurpose chair which resulted from different ergonomic studies.**
*Grid: 10×10 cm (68, 69).*

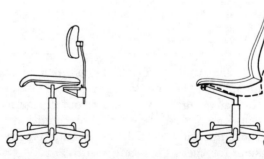

**Figure 71 A traditional office chair with an adjustable backrest (left) and a tiltable chair with a high backrest (right).**
*The tiltable chair was freely moveable from 2° forward to 14° backward and could not be locked at any desired inclination.*

from sliding forward and by a high backrest with a lumbar pad fixed at a height between 10 and 20 cm above the seat. Hünting and Grandjean (91) studied chairs with high backrests, both in the laboratory and under practical working conditions, recording sitting habits and reports of physical discomfort in different parts of the body. A tiltable chair and a similar model with a fixed seat were compared with a traditional type of chair fitted with an adjustable but short backrest. The latter and the tiltable chair are illustrated in Figure 71.

The subjects were doing their normal work while using each of the three chairs for two weeks. The most interesting results were related to the reported preferences, as shown in Figure 72.

The survey indicated quite clearly that the office workers favoured the two types of chair with the high backrest. This confirms the view expressed earlier that a high backrest is preferable for office work as most employees often desire to lean back. It is obvious that a high backrest will be more effective in supporting the weight of the trunk than a chair with a small backrest. It follows that *any office working place offering the possibility of leaning back — all the time or only occasionally — should be provided with a high backrest*. It must be pointed out here that the experimental tiltable chair could not be fixed at the desired inclination; thus it did not provide enough support for the whole body. This was criticized by many subjects and leads to the conclusion that tiltable chairs

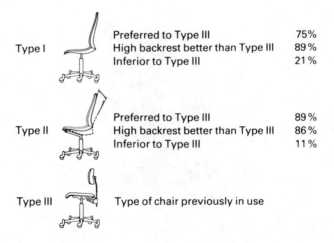

**Figure 72 Comparative assessment of three experimental chairs which were used by 66 office employees over a period of two weeks for each chair.**
*Type I = fixed moulded chair with high backrest. Type II = moulded chair with tilts of 2° forward and 14° backward, freely movable with high backrest. Type III = standard office chair with a rather short but adjustable backrest.*

or chairs with adjustable backrest inclinations should be fitted with a mechanism allowing the subject to fix the desired degree of inclination.

General experience as well as a number of studies have yielded the following 'golden rules' for office chairs:

*Golden rules for office chair equipment*

1. *Office chairs must be adapted to both traditional office work and the modern equipment of information technology, especially to jobs at VDT workstations.*
2. *Office chairs must be conceived for a forward and reclined sitting posture.* See Figure 73.
3. *The backrest should have an adjustable inclination*, and it should be possible to lock the backrest at any desired inclination.
4. *A backrest height of 48–52 cm vertically above the seat surface is an ergonomic necessity today.* The upper part of the backrest should be slightly concave. A width of 32–36 cm for the backrest is advisable. It may alternatively be concave in all horizontal planes with a radius of 40–50 cm.
5. *The backrest must have a well formed lumbar pad*, which should offer good support to the lumbar spine between the third vertebra and the sacrum, e.g., at a height of 10–20 cm above the lowest point of the seat surface. These recommendations are illustrated in Figure 74.
6. *The seat should measure 40–45 cm across and 38–42 cm from back to front.* A slight hollow in the seat, with the front edge turned upwards about 4–6° will prevent the buttocks from sliding forward. A light padding of foam rubber

**Figure 73  An office chair must be conceived for a forward as well as backward inclined sitting posture.**
*The lumbar spine must get proper support from the backrest in both sitting postures.*

# Ergonomic design of VDT workstations

The seat at a VDT workstation

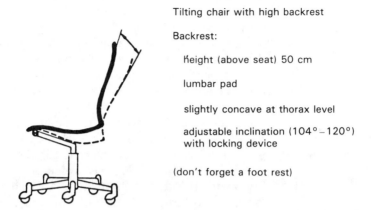

Tilting chair with high backrest

Backrest:

   Height (above seat) 50 cm

   lumbar pad

   slightly concave at thorax level

   adjustable inclination (104°–120°) with locking device

(don't forget a foot rest)

**Figure 74** **Recommendations for the backrest of a well designed office chair.**

2 cm thick, covered with non-slip, permeable material is a great aid to comfort.
7. *Foot rests are important*, so that small people can avoid sitting with hanging feet.
8. *An office chair must fulfil all requirements of a modern seat*: adjustable height (38–54 cm), swivel, rounded front edge of the seat surface, castors or glides, 5-arm base and user-friendly controls. The most important dimensions for a seat and working desk are shown in Figure 75.

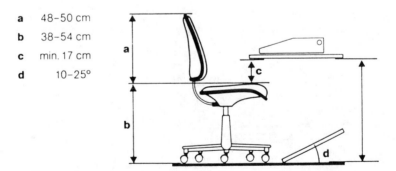

a   48–50 cm
b   38–54 cm
c   min. 17 cm
d   10–25°

**Figure 75** **Recommended dimensions for the design of the seat and working desk.**

*The backrest is crucial for an office chair*

Orthopaedics as well as ergonomics recommend frequent or at least occasional changes of position from leaning forward to leaning back and vice versa. This calls for a 'dynamic' chair which allows easy changes of the sitting posture. It is obvious that an adjustable back-rest is crucial for such a 'dynamic' chair. Such chairs are available with a tiltable shell or with a back-rest tilting independently of the seat surface which, for its part, can remain in a horizontal position or be tilted backwards with increasing backrest declination. With the tiltable shell the angle between backrest and seat surface remains the same in all positions. A drawback of this chair is the elevation of the knees with full backrest declination. With the independently tilting backrest a simultaneous inclining of the seat surface is advisable to prevent forward sliding.

*The "Syntop" chair*

The most sophisticated office chair, especially adapted to VDT workstations, is the "Syntop" model which was presented by Hort at the 'Ergo-design 84' conference (90) and is produced by Giroflex Entwicklungs A.G. of Koblenz, Switzerland. The development of this chair is based on the fact that the lumbar support of a backrest moves about 4.5 cm upwards when the inclination is increased from 90° to 105°. This corresponds to almost a whole lumbar vertebral segment. A consequence of a declined backrest is therefore that the lumbar support is no longer at the correct level but too high. The main characteristic of the new chair is that the backrest descends as its declination increases. This is shown in Figure 76.

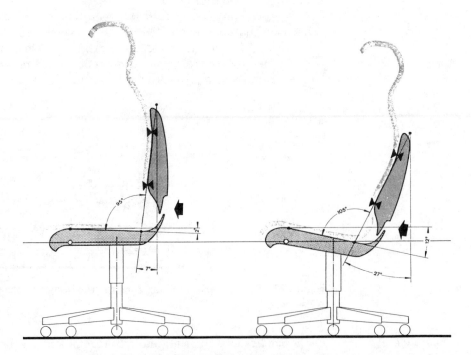

**Figure 76 A recent office chair design in which the backrest moves down with increasing declination.**
*This mechanism allows the back to get adequate support at the correct level for any backrest declination, as indicated by the small arrows. The large arrows show the way the backrest descends with increasing declination.*

The "Syntop" chair is an interesting example of applying ergonomics to the design of a chair.

The main objection to a good office chair with an adjustable backrest inclination is of course its cost. But one should bear in mind that the life span of a well constructed chair is about 10 years, or about 2000 working days. The price for a good chair which reduces physical discomfort and promotes well-being is, at a few cents per day, certainly a good investment!

## 7.7. VDT workstation design: preferred settings and their effects

*Risk of constrained postures*

At traditional office workplaces the risk of constrained postures is low, since workers perform a variety of activities. Given such working conditions no employee will mind or complain if the design of the workplace is not optimal. However, unsuitable settings will be of crucial importance for people who adopt a constrained posture when working with VDTs or other office machines. Every inadequacy of design or dimension will, in the long run, generate static efforts associated with muscle fatigue, stiffness and pains in the neck—shoulders—arm—hand region. This is the reason why adjustable VDT workstations have appeared on the market in the last few years, mainly with the argument that a workstation should be adaptable to the different anthropometric features of employees.

Several experiments have been carried out under laboratory conditions with the aim of assessing the preferred settings. Since there has been no agreement about the results it is necessary to give a brief description of the different designs and conditions of the experiments.

*Laboratory experiments*
*Miller and Suther*

1. Miller and Suther (142) used 22 male and 15 female subjects; 10 operators belonged to the 5th percentile (very short people), 11 to the 50th and 16 to the 95th percentile (very tall people) of physical stature. The average body height was 171 cm. To assess preferred settings the subjects were set the task of typing one page. It was not necessary to look at the screen. No information on the distance between the keyboard and table edge or the possibility of resting forearms and wrists is given.

*Brown and Schaum*

2. Brown and Schaum (24) experimented with 100 subjects: 40 females (mean body height, $\bar{x} = 164$ cm) and 60 males (mean body height, $\bar{x} = 179$ cm). The subjects played a word guessing game. After each of the six words they had guessed, the subjects could adjust the VDT components. There was no typing involved and there was no obligation

to look at the screen. The ranges of adjustability of the equipment were insufficient, as 17 subjects set the screen at the upper limit of 81 cm above the floor and 9 subjects set the keyboard (home row) at the lower limit of 71 cm above floor.

Grandjean *et al.*
3. Grandjean *et al.* (72) tested 30 trained female typists. The group had a normal distribution of body statures, exceeding the mean body height of European women by 5 cm (=166 cm). Thirteen subjects wore glasses. Only two subjects reported pains in neck and shoulders during the last few weeks. The home row of the keyboard was 8 cm above the desk level. A support for forearms and wrists was provided (see Figure 76). The chair had a high backrest with an adjustable inclination. The subjects typed a text of five lines on the screen and afterwards copied the same text again and again for 10 minutes. The preferred dimensions were assessed before, during and after the 10 minute typing test. After that the subjects had to repeat the typing tasks with imposed settings.

Cushman
4. Cushman (40) tested 20 experienced female VDT operators who entered text from paper copy for 50 minutes. Their average stature was 164 cm with a standard deviation (S.D.) of ± 8 cm. The subjects performed the task at five different keyboard heights (from 70 to 86 cm above floor) for 10 minutes each. Keying rate and error data as well as subjective judgements were obtained for all five test conditions. The keyboard was 7 cm high and movable. An adjustable chair with a fixed backrest inclination was provided. There was no hand rest in front of the keyboard.

Rubin and Marshall
5. Rubin and Marshall (172) tested 25 men and 25 women aged between 17 and 73 at three different VDT workstation positions. Five groups were formed, each consisting of five males and five females, corresponding in height to the 5th, 25th, 50th, 75th and 95th percentile of the British civilian population. All subjects were naive users of keyboards and VDTs, so they had to glance frequently from the screen to the keyboard to ensure correct key selection. The three positions are defined as follows: a "standard position", corresponding to dimensions that might be found in a typical office; a "users preferred position", taking into account the preferred settings of the subjects; and an "ergonomist determined position" which meets the recognized human factors recommendations. Each experiment lasted 10–15 minutes.

Weber *et al.*
6. Weber *et al.* (206) recorded the EMG of the trapezius muscle at the preferred keyboard height as well as 5 cm

above and below it. Each condition was tested with and without forearm—wrist support. Furthermore, the pressure load of the forearms, wrists and hands on the support and the keyboard was recorded. Twenty trained subjects had to imitate a VDT job by operating the keyboard and looking alternately at the source document and screen. Each experiment lasted 10 minutes.

*Preferred settings under laboratory conditions*

The preferred settings obtained in these six laboratory studies are assembled in Table 24.

**Table 24 Preferred settings of adjustable VDT workstations from six laboratory experiments.**

|  |  | Reference |  |  |  |  |  |
|---|---|---|---|---|---|---|---|
|  |  | (24) | (40) | (72) | (142) | (172) | (206) |
| Keyboard height[a] | $\bar{x}$ (cm) | 74 | 74–78[e] | 77 | 71 | 70·5 | 78[f] |
|  | range | 72–84 | — | 1–84 | 64–80 | — | 74–84 |
| Screen height[b] | $\bar{x}$ (cm) | 100 | — | 109 | 92 | 86.7 | 97 |
|  | range | 88–108 | — | 94–118 | 78–106 | — | 85–108 |
| Screen angle[c] | degrees | 10° | — | 0° | 3° | — | 11° |
|  | range | 3–17° | — | 0–16° | 0–7° | — | 0–21° |
| Screen distance[d] | $\bar{x}$ (cm) | 52 | — | 65 | — | — | 71 |
|  | range | 44–46 | — | 47–94 | — | — | 60–96 |
| Seat height | $\bar{x}$ (cm) | 50 | — | 47 | 41 | 41.8 | 47 |
|  | range | 44–52 | — | 43–51 | 32–49 | — | 43–55 |

$\bar{x}$ = mean values; [a] = home row height above floor; [b] = centre of the screen above floor; [c] = upward tilted screens related to a vertical line; [d] = screen centre to table edge; [e] = settings with best subjective ratings, highest keying performances and lowest error rates; [f] = with wrist support.

As mentioned earlier, the experimental conditions of the six studies differed greatly from each other (simulated VDT work versus other test activities, different choice of stature distribution, "with" versus "without" wrist support, trained versus naive subjects). The disagreement about the results of Table 24 is therefore not surprising. Nevertheless a few tendencies do emerge from these results.

In all studies the range of preferred settings is rather wide. Taking into account all extreme figures the following ranges can be observed:

1. Keyboard height: 64–84 cm
2. Screen height: 78–118 cm
3. Screen distance from table edge: 44–96cm
4. Screen angle: 0–21°
5. Seat height: 32–55 cm

*Many prefer keyboard heights above the recommended levels*

The heights of the keyboard are not in accordance with the usual ergonomic recommendations which are mainly based on anthropometric considerations. The ranges reveal that a large number of operators prefer keyboard heights above the recommended level of between 72 and 75 cm. Cushman (40) obtained preferred heights which were 5–10 cm above elbow level, which is contrary to the information given in ergonomic textbooks, which recommend elbow height.

The preferred values for the screen level are in general also higher than recommended. About 50% of the subjects fixed the screen centre at levels exceeding 95 cm above the floor. That means that many operators prefer a nearly horizontal line of sight when looking at the screen or a slightly downward visual angle. This seems to correspond well to the 'normal' line of sight.

All six laboratory studies have one major drawback in common: the experiments were carried out only over a short period of 10 minutes or even less. It is highly doubtful whether subjects engaged in a short-term experiment will have the same postures or prefer the same settings as those working at a VDT workstation for months or years.

*Feelings of relaxation determine preferred settings*

In spite of these shortcomings the laboratory studies disclosed other interesting results which shall be discussed here briefly. The experiments with preferred and imposed settings (72) revealed an increase in physical discomfort in the neck–shoulder–arm region under the conditions of imposed keyboard height and screen distance. These results, compared with those of Cushman (40), lead to the conclusion that the subjects are guided by a feeling of relaxation when assessing the preferred workstation settings, which are, in turn, associated with high keying performance and low error rates.

*Wrist supports*

Twenty of the 30 subjects preferred a keyboard with wrist support and 24 subjects claimed that the wrist support does not impede typing activities (72). Weber et al. (206) examined the effects of wrist support in a more systematic way. The pressure load exerted on the support remained suprisingly constant over the 10 minute typing periods. Without wrist support the pressure load on the keys was nearly zero; when working with support it ranged between 15 and 35 N (ca. 1.5–3.5 kp) on average and increased significantly with higher keyboards. These results are reported in Figure 77. They confirm the findings of Occhipinti et al. (163) shown in Figure 60.

The reader is reminded here that resting upper limbs on a support markedly reduces disc pressure in the lumbar spine.

Under each working condition with wrist support there was a significant negative correlation between EMG figures and exerted pressure load. *This implies that the more the arms and hands rest on the support, the lower is the electrical activity in the trapezius muscle.* The results are shown in Figure 78. A comparison of the experiments with and without wrist sup-

*Ergonomic design of VDT workstations* 139

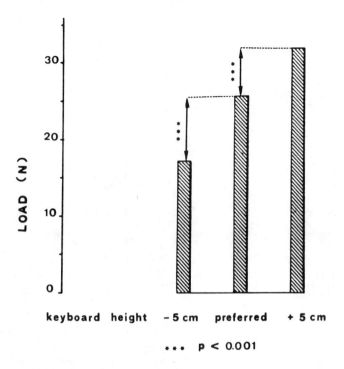

Figure 77 Mean pressure load on wrist support and keyboard during three experimental conditions.
*Means of 10 min periods for 20 trained subjects. According to Weber et al. (206).*

port showed that in the former the EMG activity of the trapezius muscle was always lower, independent of keyboard height. At the end of the experiment 12 of the 20 subjects preferred a keyboard with wrist support.

The results of the above six studies were not confirmed by the other experimental approach to keyboard operation carried out by Life and Pheasant (132). A preceding survey of 14 typists had shown that they worked with high keyboard levels (ca. 80 cm) with a mean vertical elbow–keyboard distance of 8.7 cm. The authors assumed that the reported complaints of discomfort were related to the dimensions of the workplace. For the ensuing experiments the authors engaged 12 skilled female typists who performed a typing task first at elbow level and then 5, 10 and 20 cm above elbow height. (No information is given about the existence or the use of a wrist support!) Six subjects read the copy script from a stand, whereas for the other six the copy was laid flat on the desk. The subjects did not look at the screen during the task. The performance was virtually unaffected when the keyboard was raised from elbow level to 20 cm above elbow level, but there

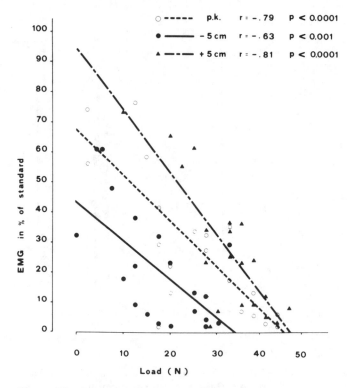

Figure 78 Relationship between EMG activity and exerted pressure load for each experimental condition with wrist support.
r = Pearson correlation coefficient; p.k. = preferred keyboard height; ±5 cm = below/above the preferred keyboard height. Means of 10 min working periods for each of the 20 subjects. According to Weber et al. (206).

was a consistent increase in the torque in the shoulder as the working height was increased. This means that a higher keyboard level requires more static activity by the shoulder muscles to support the weight of the upper limb. The torque at the neck (C7 articulation) slightly decreased when the keyboard level was raised. The six subjects with the script in the stand complained about a significant increase of discomfort as the keyboard level was raised. This effect was less pronounced in the subjects working with the laid out script. The strongest discomfort was felt in the forearms, arms and shoulders. The authors conclude that the home row of the keyboard should be approximately at the elbow height of the operator to reduce shoulder load. This conclusion might be valid for typists who adopt an upright trunk posture and do not rest the wrists on a support. This study by Life and Pheasant (132) demonstrates how delicate laboratory experiments can be and how dangerous it is to draw general conclusions. Who knows what the latter experiments would have revealed if the subjects had been given chairs with pro-

*Serving customers requires other postures*

per backrests, suitable supports for the wrists and a task requiring visual contact with the screen? These objections might well be the reason for the disagreement among all the above mentioned studies.

Another study, carried out by Launis (127), shows that quite a few jobs with VDTs require very peculiar postures associated with unusual dimensions of workstations. The optimum level and acceptable range of seat and table heights were assessed for 30 female operators engaged in the computerized selling of railway tickets. The test was carried out at a workstation, with the subjects simulating their job of selling tickets with computer-aided equipment. The operators were constantly focused on the customers so that they were able to "read from the lips". Under noisy conditions they used to lean forward and stretch themselves as far as possible. The preferred levels of the seats were 44 cm (range 36–52 cm) high and the optimum table heights were between 17 and 26 cm above seat level, and on average 1.4 cm lower than the elbow height. This particular working situation, with a readiness to serve customers, decisively determined these results.

*Preferred settings of VDT workstations in offices*

An extensive study on postures and preferred settings of adjustable VDT workstations during subjects' usual working activities was carried out by Grandjean *et al.* (73). The experiments were conducted on 68 operators (48 females and 20 males aged on average 28 years) in four companies: 45 subjects had a conversational job in an airline company, 17

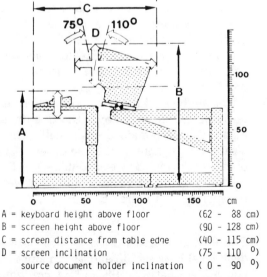

A = keyboard height above floor (62 – 88 cm)
B = screen height above floor (90 – 128 cm)
C = screen distance from table edge (40 – 115 cm)
D = screen inclination (75 – 110 °)
   source document holder inclination ( 0 – 90 °)

**Figure 79 The adjustable VDT workstation with the ranges of adjustability, used in a field study during subjects' usual working activities.**

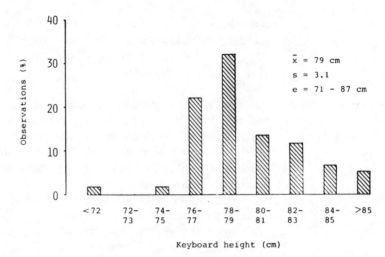

Figure 80  Preferred keyboard heights of 59 VDT operators (236 observations) while performing their usual daily jobs.
Keyboard height = home row above floor. $\bar{x}$ = mean value; s = standard deviation; e = range.

subjects had primarily data entry activities in two banks and 6 subjects were engaged in word processing operations. Each subject used the adjustable workstation shown in Figure 79 for one week.

The keyboard height was 8 cm above desk level, and a chair was provided with a high backrest and an adjustable inclination. For the first two days a forearm—wrist support was used; on the following two days the subjects operated the keyboard without support and on the last day they were given the option to use it or not. Document holders were provided as an optional supporting device for each subject. The preferred settings and postures were assessed and determined every day.

The analysis of the results of preferred settings disclosed no noticeable differences over the five days. In other words, the mean values remained practically the same for the whole week, independent of the use of wrist support. Thus the data obtained during the week could be put together for evaluation.

The frequency distribution of all preferred keyboard heights is reported in Figure 80.

*Range of adjustability for keyboard desk*

The 95% confidence interval lies between 73 and 85 cm. A desk level between 63 and 79 cm suits a keyboard height of 8 cm, a level between 68 and 84 cm would be adequate for a keyboard height of 3 cm. Assuming the 95% confidence interval *the range for the adjustability of desk levels lies between*

**Table 25** Preferred VDT workstation settings and eye levels during habitual working activities.

| Adjustable dimension | $n_1$ | $n_2$ | Mean | Range |
|---|---|---|---|---|
| Seat height (cm) | 58 | 232 | 48 | 43–57 |
| Keyboard height above floor (cm) | 59 | 236 | 79 | 71–87 |
| Screen height above floor (cm) | 59 | 236 | 103 | 92–116 |
| Visual down angle (eye to screen centre, °) | 56 | 224 | −9 | +2−26 |
| Visual distance (eye to screen centre, cm) | 59 | 236 | 76 | 61–93 |
| Screen upward inclination (°) | 59 | 236 | 94 | 88–103 |
| Eye level above floor (cm) | 65 | 65 | 115 | 107–127 |

$n_1$ = number of subjects; $n_2$ = number of observations; visual down angles and screen inclination are related to the horizontal plane.

65 and 82 cm. This seems to be a reasonable recommendation for workstation manufacturers.

The results obtained in this field study reveal slightly higher levels of keyboards than those obtained in comparable laboratory studies, shown in Table 24.

It is assumed that in short-term experiments subjects are less relaxed, sit more upright and try to keep the elbows low and the forearms in a horizontal position, thus giving preference to a slightly lower keyboard height.

*Preferred settings of screen*

All the results of preferred settings are assembled in Table 25. The preferred screen heights and screen inclinations are in some cases influenced by the attempt of operators to reduce reflections. In fact, many operators reported less annoyance by reflections if they could adjust the screen.

The capital letters on the screen were 3.4 mm high, corresponding to a comfortable visual distance of 68 cm. At the adjustable VDT workstation the operators tended to choose greater viewing distances; 75% of them had visual distances of between 71 and 93 cm. No explanation can be found for this preference.

The visual down angles correspond well to the 'normal' line of sight, discussed earlier in Section 7.5, but they are not in agreement with those authors who recommend down angles of 38° or more (87, 130). VDT operators obviously prefer slightly declined visual down angles of 0 to 15° (=90% confidence).

The calculation of Pearson correlation coefficients between anthropometric data and preferred settings revealed only poor relationships: between eye levels and screen heights $r=0.25$ ($p=0.03$) and between body length and keyboard heights $r=0.13$ (not significant). Some of the laboratory studies revealed similar results of poor or no relationships. It can be concluded, therefore, that *the preferred settings of VDT workstations are not greatly influenced by anthropometric factors.*

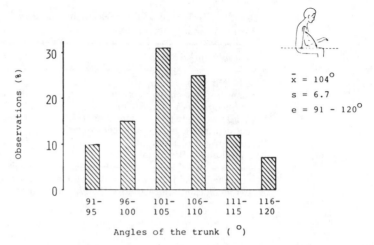

Figure 81 Trunk postures of 59 VDT operators (236 observations) while performing their usual daily jobs.
*Trunk posture is assessed as the angle of a line between the hip and shoulder points to the horizontal.* $\bar{x}=$ *mean value.* s=*standard deviation.* e=*range.*

*Most operators lean back...*

The most striking result of this field study concerns the postures associated with the preferred settings. The operators moved only very occasionally and did not noticeably change the main postural elements which are obviously determined by the position of the keyboard and screen. Figure 81 shows the distribution of determined trunk postures expressed as angles of the line "shoulder articulation to trochanter" to the horizontal plane.

The trunk inclinations approximate to a normal distribution. The majority of subjects prefer trunk inclinations of between 100 and 110° while only 10% demonstrated an upright trunk posture. Figure 82 illustrates the mean and the range of observed trunk postures. It is obvious that the majority of operators lean back. This is the basis for all other adopted postural elements: the upper arms are kept higher and the elbow angles slightly opened. The mean figures for preferred trunk–arm positions are listed in Figure 83.

*...and hold arms and hands slightly raised*

It must be pointed out here that about 80% of the subjects do rest their forearms or wrists if a proper support is available. If no special support is provided, about 50% of the subjects rest their forearms and wrists on the desk surface in front of the 8 cm high keyboard.

The observed postures are not due to the experimental workstation, for the measurements carried out at the previous workstations had already revealed nearly the same trunk and arm inclinations.

*Ergonomic design of VDT workstations*

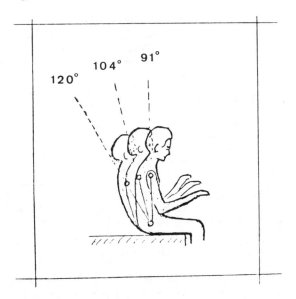

**Figure 82 Mean and range of observed trunk postures of 59 operators.**
*Trunk posture is assessed as the angle of a line between the hip and shoulder joints to the horizontal.*

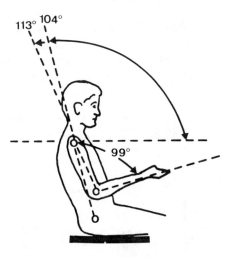

**Figure 83 The 'average posture' of VDT operators at workstations with preferred settings.**

This study confirms a general impression one gets when observing the sitting posture of many VDT operators in offices: most of them lean back and often stretch out the legs. They seem to put up with having to bend the head forward and lift their arms. In fact, *many VDT operators in offices disclose postures very similar to those of car drivers.* This is

**Table 26** Means ($\bar{x}$), standard deviations (S.D.) and ranges of postural measurements obtained from VDT operators during their daily work at workstations with preferred settings.

| Postural element | $\bar{x}$ | S.D. | Range |
|---|---|---|---|
| Trunk inclination (°) | 104 | 6.7 | 91–120 |
| Head inclination[a] (°) | 51 | 6.1 | 34–65 |
| Upper arm flexion[b] (°) | 113 | 10.4 | 91–140 |
| Upper arm abduction[c] (°) | 22 | 7.7 | 11–44 |
| Elbow angle (°) | 99 | 12.3 | 75–125 |
| Lateral abduction of hands[d] (°) | 9 | 5.5 | 0–20 |
| Acromion–home row distance (cm) | 51 | 5.0 | 42–62 |

59 operators, 236 observations.
[a] Angle C7–earhole–vertical; [b] see Figure 82; [c] abduction = lateral raising of upper arm; [d] see Figure 56

understandable: who would like to adopt an upright trunk posture when driving a car for hours?

The results of all the measured postural elements, expressed as mean values and ranges, are reported in Table 26.

*Preferred settings and physical discomfort*

The VDT operators completed a questionnaire about feelings of relaxation and physical discomfort, once at the previous workstation and twice at the adjustable workstation with preferred settings. An index was calculated from the answers 'relaxed', 'tense' and 'impaired' for each of the involved parts of the body (neck, shoulders, back, forearms and wrists). In Figure 84 the mean indices of complaints from the previous

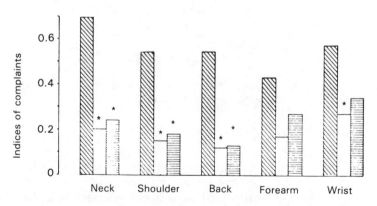

Figure 84 Mean indices of complaints at the original workstation and the redesigned workstation with preferred settings.
$0 = relaxed;\ 1 = tense;\ 3 = impaired.$ *$p \leqslant 0.05$.

*Discomfort is reduced with preferred settings*

workstation are compared with those reported on the second and fourth days at the redesigned, adjustable workstation. An index of less than 0.5 means that the majority of the subjects rated their postures as relaxed; an index of more than 0.5 means that many subjects indicated that their muscles were tense or even that they experienced impairments. From Figure 84 it is obvious that the indices were distinctly higher at the previous workstation than with the preferred settings. A chi-squared analysis showed significant differences between the previous and adjustable workstation for the neck, shoulders and back. It must be pointed out that at the previous workstation subjects sat on traditional office chairs with relatively small backrests. At the adjustable workstation, however, they were provided with particularly suitable office chairs, featuring high backrests with adjustable inclinations, which allowed the whole back to relax. It is therefore reasonable to assume that the decrease of physical discomfort reported at the adjustable workstation was due to both the preferred settings and the proper chairs.

*Results confirmed by the Shute and Starr study*

These results were confirmed by Shute and Starr whose study (182) was already mentioned in Section 7.3. In a first field study telephone operators were provided with an adjustable VDT table and in a second study with an additional conveniently adjusted chair. Subjects used the adjustable workstation for several weeks while doing their normal work. The previous VDT table had a fixed height of 68.6 cm and the screen was 40 cm above the table. The previous chair was difficult to adjust and had an unsuitable backrest. The main difference of the new, advanced chair was mainly its easy adjustability. The results revealed a reduction of discomfort when either conventional component was replaced by an advanced component. But the reduction in physical discomfort was far greater when the advanced table was used together with the advanced chair. The authors concluded that the benefit of an advanced table can only be fully realized if it is used in combination with an advanced chair.

*Preferred settings at CAD workstations*

Van der Heiden and Krueger (83) examined the use and acceptance of an adjustable workstation for CAD operations. Height and inclination of work surface as well as height, inclination, rotation and distance of the monitor could be adjusted with the aid of motorized devices. To study the use of the adjustment fixtures the settings were continuously monitored for one week. The majority of operators had more than six weeks experience in using the adjustable workstation. In the test week eight women and three men were studied during their normal CAD work, mechanical design. A total of 67 CAD work sessions were monitored. Questionnaires and preferred settings were obtained from eleven

**Table 27** Preferred settings of an adjustable CAD workstation by 15 operators.

| Parameter | $\bar{x}$ | e |
|---|---|---|
| Seat height (cm) | 54 | 50–57 |
| Work surface height (cm) | 73 | 70–80 |
| Monitor centre above floor (cm) | 113 | 107–115 |
| Monitor visual distance (cm) | 70 | 59–78 |
| Work surface tilt[a] (°) | 8.6 | 2–13 |
| Monitor tilt[a] (°) | −7.7 | −15– +1 |

$\bar{x}$ = mean values; e = range; [a] negative tilt = a forward monitor inclination (top of the screen toward the operator).

female and four male operators. Of a total of 166 registered adjustments 142 (=86%) were made at the beginning of a work session and 24 (=14%) were readjustments. Short operators used the adjustment device more frequently than tall people. Furthermore, operators who had not received specific instructions adjusted less frequently than others who had been given such instructions. The preferred settings are presented in Table 27.

The mean seat height of 54 cm is quite unusual, but might be related to the group of rather tall operators with a body length between 158 and 185 cm. Another striking result is the forward tilting of the monitor with a preferred mean angle of −8°. The operators claimed that with this setting reflections resulting from windows behind them could be avoided. For that reason most operators preferred a relatively high setting of the monitor. All other preferred dimensions are similar to those of the VDT operators as shown in Table 25.

*Wishful thinking of standards versus operators' instinctive behaviour*

Let us come back to the backward declination of the trunk observed at VDT workstations. This leaning back does not correspond at all to the commonly published and recommended values for postures (13, 27, 45, 203). Figure 85 illustrates the great gap between 'wishful thinking' (recommendations) and actual postures. An important question suggests itself here. Is the upright posture healthy and therefore recommendable or is the relaxed position with the reclining trunk to be preferred? As already mentioned in Section 7.5, by increasing the backrest inclination from 90° to 120° a significant decrease of discal load and muscle strain is achieved (4), as shown in Figure 66. These orthopaedic studies (4) suggested that *resting the back on a sloping backrest transfers a relevant portion of the trunk weight to the backrest and reduces strain on discs and muscles more than it does when sitting straight and upright. It is therefore concluded that VDT operators instinctively do the right thing when they prefer a reclined sitting posture and ignore the recommended upright trunk position.*

# Ergonomic design of VDT workstations

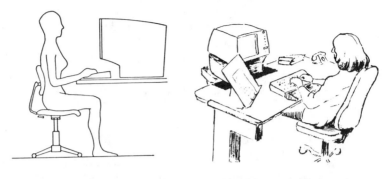

Wishful thinking        Preferred body posture

**Figure 85 Recommended and actual postures at office VDT workstations.**
*Left: the upright trunk posture with elbows down and forearms almost horizontal, postulated in many brochures and standard works. Right: the actual posture most commonly observed at VDT workstations resembles the posture of a car driver.*

One restriction must be made here: some special work situations (such as manual work requiring freedom of movement or physical effort) might call for an upright trunk position with elbows down and forearms horizontal. As discussed in Section 7.2, the old mechanical typewriters requiring key forces of several hundred grams were more easily operated in such a posture. But the advances in electronic keyboard technology today permit very rapid keying with low key forces of 40–80 g and key displacements of 3–5 mm. Modern keyboards are mainly operated by finger movements with hardly any assistance of the forearms. These conditions might to some extent explain why VDT operators in offices prefer to lean back, keep the upper arms slightly forward with the wrists on a support (which can be the desk itself) and adjust the keyboard to a rather high level.

*Guidelines for the design of VDT workstations*

From the study on preferred settings the following guidelines for the design of a VDT workstation can be proposed:

*The furniture should, in principle, be as flexible as possible. A proper VDT workstation should be adjustable in the following dimensions:*

*Keyboard height (floor to home row)*    70–85 cm
*Screen centre above floor*    90–115 cm
*Screen inclination to horizontal*    88–105 cm
*Keyboard (home row) to table edge*    10–26 cm
*Screen distance to table edge*    50–75 cm

*A VDT workstation where the keyboard height is not adjustable and where the screen height and operator–screen distance are not adjustable is not suitable for continuous work.*

*The controls for adjusting the dimensions should be easy to handle, particularly at shift-work workstations where a series of operators work throughout the day.*

*At knee level the distance between the front table edge and the back wall should not be less than 60 cm and at least 80 cm at the level of the feet.*

## 7.8. The VDT keyboard

*Parallel rows require unnatural positions of hands*

The keyboard for typing letters was invented in 1868, as a mechanical device with 4 parallel rows of keys. To operate these keys rapidly, the typist must hold the hands parallel to the rows. This requires an unnatural position of the wrists and hands, characterized by an inward rotation of the forearms and wrists and a lateral (ulnar) abduction of the hands. These constrained postures often cause physical discomfort and in some cases even inflammation of tendons or tendon sheaths in the forearms of keyboard operators. Figure 86 illustrates the constrained posture of the wrists and hands at a keyboard.

In Section 7.3, Figure 58 showed the incidence of medical findings in the right forearm of VDT operators who exhibited a strong lateral abduction of the right hand.

In time the electric typewriter replaced the mechanical typewriter and more recently electronic typewriters have been introduced.

The mechanical resistance of keys has been much reduced and the operation of the keyboard made easier, but the unnatural position of wrists and hands has remained.

**Figure 86 Position of wrists and hands operating a traditional keyboard.**
*The parallel position to the rows requires an inward rotation of the forearms and wrists and a sideward twisting (lateral abduction) of the hands.*

*Flat keyboards at VDTs*

At VDT workstations the typing activity is similar to the traditional operation of typewriters. There are some slight differences, though. Firstly, the number of keys has increased, with specially arranged numerical keys and several functional keys for operating the computer. Secondly, in conversational jobs operators must frequently wait for the response of the computer. This response time can last from one to several seconds. According to the Swedish study of Johansson and Aronsson (99) response times of more than 5 s were found annoying and stressful by operators. During these unwanted pauses operators like to rest forearms and wrists on suitable supports. This has induced some VDT designers to develop flat keyboards which allow operators to rest their forearms and wrists on the desk. For this reason many ergonomists nowadays recommend *a flat keyboard with a home row not higher than 3 cm above the desk and the opportunity for the operator to move the keyboard on the desk according to his/her needs.*

The next step should be an ergonomic design of the keyboard to avoid the constrained hand posture by reducing or eliminating the inward rotation and lateral abduction of the hands and by a proper support for the forearms and wrists.

*Studies on split keyboards*

Studies along these lines were carried out as early as 1926 by Klockenberg (103) and in 1965 by Kroemer (112, 113), who proposed splitting the keyboard into two parts and arranging them in such a way that the hands could be kept in a more natural position. Kroemer's halved keyboards had an opening angle of 30° and a lateral declination adjustable between 0° and 90°. The opening angle is defined by two lines running through the inner board of the keys Y—H—N and T—G—B. Experiments showed that the experimental keyboard generated less painful fatigue than traditional typewriters (112, 113). Webb and Coburn (205) observed increased typing performances when using hand-configured keysets and laterally sloped keyboards.

*EMG of forearms and shoulders*

In his electromyographic study on typists Lundervolt (136) recorded increased electrical activities in the forearm muscles which are responsible for the inward rotation of the hands.

Recently Zipp et al. (214) studied the electrical activity of various muscles in the shoulder—arm area in relation to hand—arm postures according to the characteristics of keyboard operations. With increasing lateral abduction of the hands from a neutral position, an increase in the electrical activity of the muscles involved was recorded. A split keyboard, as proposed by Kroemer, resulted in decreased electrical activities in the arm—shoulder area, obtained already with lateral keyboard inclinations of 10–30°. The same result was observed when the angle between the two keyboard halves was opened. The authors concluded that the static muscle load in the arm—shoulder area is significantly reduced with such a keyboard design.

*Experiments with halved keyboards*

Following this line of research, Grandjean *et al.* (71), Hünting *et al.* (94) and Nakaseko *et al.* (150) developed an adjustable model of a split keyboard and studied preferred settings of opening angles, lateral sloping and distances between the split keyboards on 51 subjects. Typing with the split keyboard with preferred settings decreased the lateral abduction of the hands shown in Figure 87, reduced discomfort and increased feelings of being relaxed in the neck—shoulder—arm—hand region.

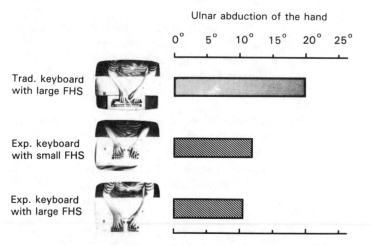

**Figure 87   Mean angles of sideward twisting (lateral abduction) of the right hand with three types of keyboards.**
*Upper: traditional typewriter with large forearm—wrist support (20 cm).*
*Middle: split keyboard with an opening angle of 25° and a lateral slope of 10°. Small forearm—wrist support (10 cm).*
*Lower: the same split keyboard but with a large forearm—wrist support (20 cm).*
*FHS: forearm-hand support.*

*Effects of a large forearm—wrist support*

The use of a large forearm—wrist support was associated with an inclined sitting posture and with an increased pressure load of forearm—wrists onto the support, reaching mean values of nearly 40 N (4 kg). According to Occhipinti *et al.* (163) such weight transfers onto the support will strongly decrease the load on the intervertebral discs (see also Figure 59). Of 51 subjects, 40 preferred the split keyboard with the following characteristics:

*Ergonomic design of VDT workstations* 153

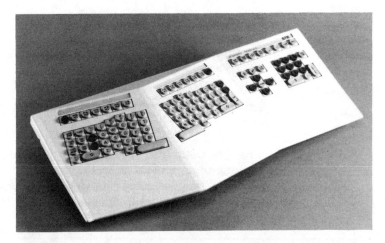

**Figure 88 A keyboard designed in accordance with ergonomic principles.**
*Two keyboard halves show an opening angle of 25° in order to avoid a sideward twisting of the hands and are provided with lateral slopes of 10° to lessen the extent of inward rotation of the forearms and wrists. According to Nakaseko et al. (150).*

*Preferred settings of split keyboards*

| | |
|---|---|
| Angle between the two half-keyboards: | 25° |
| Distance between the two half-keyboards (measured as distance between the keys 'G' and 'H'): | 9.5 cm |
| Lateral sloping of both half-keyboards: | 10° |
| A hand-configurated design of the keys | |

A prototype of such a keyboard is shown in Figure 88. A commercial model, produced by Standard Telephone and Radio AG of Zürich, was presented at the 'Ergodesign 84' conference by Buesen (26).

*Guidelines for the design of VDT keyboards*

In the last four decades, manufacturers of typewriters have greatly improved the design and the mechanical characteristics of keys. There are nearly no controversial opinions and the following guidelines are today widely accepted for the design of VDT keyboards:

| | |
|---|---|
| Keyboard height above desk (middle row) | 30 mm |
| Keyboard height (front side) | 20 mm |
| Inclination | 5–15° |
| Distance between key tops | 17–19 mm |
| Resistance of keys | 400–800 mN |
| Key displacement | 3–5 mm |

The operator should also feel when the stroke has been accepted; this is called the tactile feedback. The best feedback quality is achieved when the point of acceptance and pressure is located about halfway down the key displacement.

Finally the reader is once again reminded that the keyboard

should be movable on the desk. A good support for forearms and wrists with a depth of at least 15 cm should be provided. Recommendations for reflectances and colours of keyboards are given in Section 5.3.

## 7.9. Arrangement of work surfaces in computerized offices

The great number of different jobs associated with the varying requirements of source documents create such a variability of requirements that only very general guidelines for the arrangement and dimensions of work surfaces can be given here.

*For data-entry tasks*

The arrangement of screen, keyboard and source document should be adapted to the way and frequency these elements are used. For data-entry tasks, keyboard and source documents are used almost continuously while vision is only occasionally directed towards the screen. It is therefore recommended that the keyboard and a document holder should be positioned in front of the operator and the screen positioned to the side. Figure 89 illustrates this situation.

If the source documents are to be handled continuously (e.g., coupons or small cards), they must be arranged to the left of the keyboard within easy reach of the left hand.

Figure 89 **Recommended arrangement of screen, keyboard and document holder for a data-entry task.**
*Keyboard and source document are used continuously and should be arranged in front of the operator. The screen is viewed from time to time to verify entries and can be placed at the side.*

*For conversational tasks*

In conversational tasks (or interactive communication) the eyes mainly alternate between source documents and the screen, while both hands operate the keyboard. A frontal arrangement of all three elements is observed frequently (keyboard, source document and the screen behind), but with

# Ergonomic design of VDT workstations

large documents a compromise is necessary: they must be placed to the left of the keyboard. These alternative arrangements are also appropriate for data acquisition and word processing. Figure 90 illustrates a proper arrangement for conversational tasks.

It is obvious that all the above mentioned arrangements require great flexibility, allowing all three elements — screen, keyboard and source document — to be moved easily on the desk surface.

**Figure 90 Recommended arrangement for a conversational task.**
*If the source document is not too large it can be placed between the keyboard and the screen, otherwise it must be positioned at the side of the keyboard.*

*Work surfaces*

Another important aspect for the design of the workstation is certainly the assessment of the layout. It is necessary to know the use that is made of documents or reference manuals: their size and number, and the need to write notes either as a separate activity or in conjunction with the terminal. Furthermore, one should consider the use made of other job aids, such as the telephone, calculator, typewriter, card indexes, diaries, printers and copiers. All these needs will determine the space in the office for one VDT workstation as well as the dimension of the work surface.

*Arrangement of tables*

Several tables might be necessary to fulfill all the above mentioned needs. The tables may be placed at right angles, next to each other or at 45° to each other. The experts of the Bell Telephone Laboratories (13) report that subjects preferred the latter arrangement because it required the least chair motion to move from one section to the other. The 45° configuration also makes it possible to combine several workstations in groups with the VDTs pointing in the same direction so that all the screens can be at right angles to the windows.

*Recommended dimensions*

It is obvious that the desk surface supporting the screen, keyboard, document holder and some of the other aids must be large enough. A common general rule says that *the work surface of the main table should have a width of 120 or — even better — of 160 cm and a depth of 90 cm.*

# 8. Noise

Noise is any disturbing sound. In practice, we call it simply 'sound' when we find it not unpleasant, and 'noise' when it annoys us.

*Sound perception*

The inner ear, which contains hair cells sensitive to vibration, provides the 'interface' where sound waves are converted into nervous impulses along the auditory nerve. The actual sound perception in the integration and interpretation of these sensory impulses in the brain.

The perception of sound is not a faithful reproduction of the whole band of frequencies, which is simply 'played' in the brain. This fact is particularly obvious in people's reaction to noise, which varies greatly from person to person. What is noise to one may be music to the other. An example of varying perception is the noise of a built-in electric fan of a VDT: many operators have got accustomed to it and don't hear it, for others this continuous low noise is irritating and annoying.

*Sound pressure*

Any sudden mechanical movement causes fluctuations in the air pressure. The extent of the pressure variation is the sound pressure which determines the intensity of the acoustic sensation.

*Pitch*

The sound frequency is the number of fluctuations per second, expressed in hertz (Hz), subjectively perceived as pitch. Most kinds of noise contain a mixture of sounds of different frequencies. If high frequencies predominate, a high-pitched noise is heard, and vice versa. The range of human hearing extends from approximately 20 Hz to 20 000 Hz, whereas the range of a piano lies between 27.5 and 4186 Hz.

The physical unit of sound pressure is the micropascal ($\mu$Pa). The weakest sound that a healthy human ear can detect is about 20 $\mu$Pa. This pressure wave of 20 $\mu$Pa is so low that it causes the membrane in the inner ear to deflect by less than the diameter of a single atom! But the ear can also tolerate sound pressures up to more than one million times higher.

*The decibel*

To accommodate such a wide range with a practical scale, a logarithmic unit, the decibel (dB) was introduced.

158    *Ergonomics in Computerized Offices*

The decibel scale uses the hearing threshold of 20 µPa as a reference pressure. Each time the sound pressure in µPa is multiplied by 10, 20 dB are added to the decibel level, so that 200 µPa corresponds to 20 dB. 1 dB is the smallest change the ear can distinguish; a 6 dB increase is a doubling of the sound pressure level, although a 10 dB increase is required to make it sound twice as loud.

*Subjective loudness*

The apparent loudness of a sound depends a good deal on its pitch or frequency. Low-pitched sounds seem much quieter than high-pitched ones. This is shown very clearly by the curve of auditory threshold for different frequencies, the lowest curve in Figure 91.

Furthermore, the figure shows the curves of equal loudness, which were determined in 1933. These curves of equal subjective loudness are expressed in phons, a unit which is no longer used.

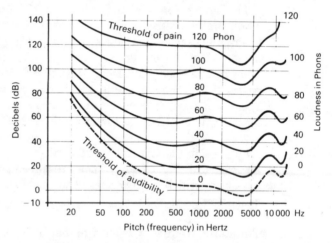

**Figure 91   Sound levels in decibels and the curves of equal subjective loudness in phons.**
*The lowest curve indicates the threshold of audibility. The diagram shows that the greatest sensitivity lies in a range between 2000 and 5000 Hz.*

*Weighted sound level*

Nowadays the so-called weighted sound level has come into use as a measure of loudness. Basically, a weighted sound level is the result of a filtering process, whereby the sound energy is filtered out at the lowest and highest frequencies, where sensitivity is less pronounced. Figure 92 sets out three weighted curves in dB, (A), (B) and (C). Today only the weighted curve (A) is widely used, since many studies have shown that noise levels measured in dB(A) permit a reliable assessment of subjective annoyance through noise.

# Noise

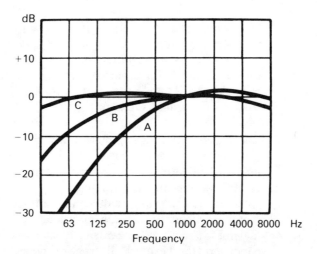

**Figure 92** Three weighted sound pressure curves in dB(A), dB(B) and dB(C), related to the frequency of sound.
*The weighted sound pressure in dB(A) is widely used today to assess annoying noise.*

*Noise load*

The noise load is the extent of noise, measured in physical units, taking account of all the acoustic factors over a given time. The sound pressure level is not the only relevant factor; the frequency of noise peaks, the noise background and other factors also contribute to the total noise load.

*The equivalent noise level $L_{eq}$*

Today, *the equivalent noise level $L_{eq}$* is the most commonly used unit to assess noise load. The $L_{eq}$ in dB(A) expresses the average level of sound energy during a given period of time. This quantity is an integration of all the sound levels which vary during this time, and so compares the disturbing effect of the fluctuating noises with a continuous noise of steady intensity. Equivalent noise levels are used to measure long-lasting noise loads, like traffice noise, but also office noise.

*The frequency of occurrence of noise levels*

Another way to measure noise is to determine the frequency of different noise levels. With this method the cumulative frequency of the most characteristic noise levels is measured with a sound level indicator and a frequency counter, operating over a given time. Commonly used units of sound measurements include: $L_{50}$ (average sound level), $L_1$ (peak noise level).

'$L_{50} = 60$ dB' means that the level of 60 dB was reached or exceeded during 50% of the relevant time.

'$L_1 = 70$ dB' means that the level of 70 dB was reached or exceeded for not more than 1% of the time.

These two levels, $L_{50}$ and $L_1$ are related to the equivalent noise level by the following approximation:

$$L_{eq} = L_{50} + 0.43(L_1 - L_{50}) = \frac{L_{50} + L_1}{2}$$

*Background noise*

When discussing noise in offices many authors also use the term 'background noise'. For some this is the average noise that continues when a machine or some other noise source has been turned off; for others it is just a not clearly defined mean noise level. In fact, background noise is in general very close to the $L_{50}$ cumulative sound level.

*Effects of noise*

The nervous impulses generated in the inner ear travel along the auditory nerve, enter the brain stem and end in the auditory sphere of the cerebral cortex, where the process of conscious hearing takes place.

In the brain stem some important side-effects occur: the incoming acoustic signals may lead to a general activation, alarming the whole organism, disturbing sleep, and reducing concentration through distraction when awake. This kind of 'alarm call' has a biologically important function in alerting a person, giving him an opportunity to interpret the noise and react appropriately to it.

Evidently, hearing has two principal functions: to convey specific information, as a basis for communication between individuals; and to activate and keep the individual awake and in a state of increased attention.

*Hearing losses*

Strong and repeated stimulation through noise can lead to a loss of hearing. Noise exposures exceeding $L_{eq}$ levels of 85 dB(A) can generate noise deafness. In offices, such noise exposures do not occur and hearing losses are not to be expected from office jobs.

*The predominant noise effects in offices are: interference with speech communication; distraction from mental activities; and annoyance.*

*Speech communication*

Where human speech is concerned, it is not enough to hear the sound pressure level of the voice. The message must also be understood, which requires a very special discriminating ability in the ear. A critical factor is the correct hearing of the consonants, which are much more difficult to distinguish than the vowels.

Speech comprehension in an office largely depends on the loudness of the voice in question, and the level of the office background noise ($L_{50}$). *Studies revealed that for good speech communication the sound pressure level of the background noise should be 10 dB or more below that of the human voice.* If the verbal communication involves an unfamiliar topic with difficult new words, then the difference between background noise level and human voice must be increased to about 20 dB.

*Voice and background noise*

The normal speaking voice, indoors, at a distance of 1 m, operates at the following sound pressure levels:

Quiet conversation 60–65 dB
Dictation 65–70 dB
Speaker at a conference 65–75 dB

If the voice has to be used frequently to dictate or to convey information to an employee, it should not exceed 65–70 dB. *If the voice is to be understood clearly and without strain, the background noise level must not exceed 55–60 dB; if the verbal communication is made more difficult, i.e., contains many unfamiliar words, then the background noise must not exceed 45–50 dB.*

*Distraction*

However, interference with speech communication is not the only noise effect in offices. Indeed, mental concentration during intellectual activities and the full attention required for skilled jobs are both disturbed by noise. Some laboratory experiments have shown that exposure to noise leads to loss of efficiency. Broadbent (23) found that a noisy situation made breaks in concentration more frequent, and thereby impaired performance in tests which called for continuous attention.

*Disturbance of intellectual activities*

It is well known that unexpected and unfamiliar stimulation through noise is also generally distracting, diverting attention from the task towards the source of noise. Such effects, in particular, interfere with intellectual activities. Furthermore, general experience has shown that people are more susceptible to noise when learning or pursuing activities which require prolonged full attention than when engaged in routine work.

*Annoyance*

It has been observed that many kinds of noise have different annoying effects which are of a more or less emotive nature. The degree of annoyance is determined by the following factors:

1. Noise level.
2. Its nature: unexpected and irregular noise is more disturbing than regular noise.
3. High-pitched noise is more annoying than low-pitched noise.
4. The kind of activity: subjects engaged in creative intellectual work will suffer more from noise than subjects doing routine work.
5. One's attitude to the source of noise: disapproval (for example, neighbours one dislikes) makes the noise more disturbing, whereas approval of the source (one's own machine, for instance) often minimizes the disturbance.
6. Earlier experience with a particular noise: the conversations among other people, for example, might arouse curiosity and therefore be very distracting.

*It can be confidently assumed that annoyance arises from the cumulative effects of noise.*

*Adaptation to noise*

The extent to which people become accustomed to noise is not yet fully understood. Experience has shown that under certain circumstances adaptation takes place, but that in other cases there may either be no adaptation or even an increased susceptibility to noise. The process of adaptation depends upon so many external and psychological factors that it is impossible, at present, to generalize.

*Outdoor noise*

The noise in offices is composed of very different sources: outdoor noise (mainly from traffic), and indoor noise, like conversations, office machines, telephone, footsteps and air conditioning.

In most streets traffic noise reaches $L_{eq}$ levels of between 60 and 70 dB(A). This can interfere with speech communication, especially in the summer when the windows are open. But nowadays many offices are conceived as open-plan offices which are well insulated against outdoor noise emissions. In such offices only the indoor noise has to be considered.

*Indoor noise...*

Noise levels to be expected in offices are listed in Table 28.

One must bear in mind that equivalent noise levels are an average value which actually lies between peak and mean noise levels (more precisely, between the cumulative levels $L_{50}$ and $L_1$). *The peak noise levels in offices are approximately 4–8 dB above the equivalent noise level* (175). These peak levels are an important factor when considering the effects of noise on speech communication.

Table 28 Noise levels commonly found in offices.

| Office | Equivalent noise level dB(A) |
|---|---|
| Very quiet, small offices and drawing offices | 40—45 |
| Large, quiet offices | 46—52 |
| Large, noisy offices | 53—60 |

*...in large offices*

Nemecek and Grandjean (155) studied noise problems in 15 large offices in 1970, at a time before VDTs had been introduced. The measured noise levels are reported in Table 29.

The results show that large offices with normal activities, including typing, are rather noisy offices. In the same survey 519 office employees were questioned about their experience and opinions. About one-third of them claimed to be "greatly disturbed" by noise and a large number (69%) mentioned "disturbance to concentration". The most surprising result of this survey is shown in Figure 93. Four hundred and eleven people who admitted to be very much or slightly disturbed by noise were questioned about the disturbing noise sources. The

# Noise

**Table 29  Equivalent noise levels ($L_{eq}$) and peak noise levels in open-plan offices.**
*According to measurements in 1970 (155, 156).*

| Open-plan office (20—120 persons) | Equivalent noise level dB(A) | Peak noise level[a] dB(A) |
|---|---|---|
| 13 offices with usual office jobs including operation of typewriters | 52—58 | 56—65 |
| 1 office with punch card machines | 62 | 65 |
| 1 office for intellectual research work | 44 | 50 |

[a] Peak noise = cumulative levels at 1% of the time.

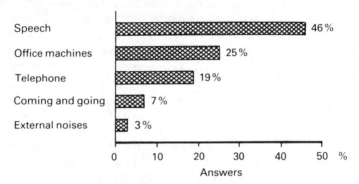

**Figure 93  Frequency distribution of replies to the question "What kinds of noise disturb you?"**
*411 employees were questioned, some mentioning more than one source, giving a total of 762 replies = 100%. According to Nemecek and Grandjean (155, 156).*

results indicate that *talk or conversation among other people was clearly the most important noise source*. Many of the subjects reported that it was the content of the conversation rather than its loudness which had the disturbing effect. Office machines, at that time mostly typewriters, were ranked second, followed by the telephone.

*Conversations are the most important noise source*

These indications were confirmed by the results of correlations calculated between noise levels and the incidence of rating noise as "very disturbing". Indeed, no correlation could be found between noise levels and disturbance. This means that the disturbance due to office noise is almost independent of the measured noise levels.

*Information content of talks is distracting*

These results led to the conclusion that *conversations of other people distract from concentration not so much through their sheer loudness as through their information content*. This observation was later confirmed in a study on offices with more traditional layout (156).

Other people's conversations which are usually not very

loud can be masked by the general background noise caused by office staff, rustling paper, walking, typing, office machines and room ventilation. One may conclude that a certain average noise level, which is covering up the conversations of others, would be appreciated by many office employees. This is why in some large offices fixtures were installed, producing a regular, constant background noise, called 'sound conditioning', to mask conversations. Keighley and Parkin (101) studied this device in a large office with 44 occupants. None of the different conditions tested was particularly successful. Noise levels between 46 and 50 dB(A) were most efficient at covering up conversations, but the acceptability of the procedure as well as its effect on face to face conversations received low ratings within this range of sound conditioning. The authors arrived at the conclusion that on the whole it cannot be proven that sound conditioning offers any real benefit nor can it be considered a universally applicable remedy for noise problems in the landscaped office. Since opinions about sound conditioning have been rather controversial, it is not yet possible to draw definite conclusions.

*Guidelines for office noise*

Based on general experience, as well as on studies, the guidelines reported in Table 30 can be proposed for office noise. It is to be recommended that the noise in offices with more than 5–10 occupants should neither be much lower nor distinctly higher than the given ranges. For offices with one or two people the recommended values can be considered to be the upper desirable limit whereas lower noise levels would certainly be more favourable. Some experience indicates that the desirable background noise is ensured in large offices of at least 100 m$^2$, with an occupancy of at least 80 persons.

The recommended range of equivalent noise levels $L_{eq}$ of 54–59 dB(A) will to some extent mask the conversations and telephone calls of others, while speech communication between two employees will remain undisturbed. This was recently confirmed by the results of a survey by Cakir *et al.* (29), who found the lowest prevalence of noise disturbances in large computerized offices with a background noise between 48 and 55 dB(A). Below and above this range the incidence of complaints was clearly higher.

Table 30  Guidelines for noise levels in open-plan offices.
*The office noise should principally be neither lower nor higher than the given ranges.*

| Noise measurements | Desirable ranges |
|---|---|
| Equivalent noise level $L_{eq}$ | 54–59 dB(A) |
| Mean noise level ($L_{50}$) (approx. background noise) | 50–55 dB(A) |
| Peak noise levels ($L_1$) | 60–65 dB(A) |

# Noise

*Large offices with VDTs*

No systematic research has been carried out on the noise in offices with VDTs. A few measurements indicate that the noise in computerized offices is approximately the same as in large offices with traditional activities, including the use of typewriters. Cakir et al. (29) measured background noise levels between 30 dB(A) and 60 dB(A) in a few large offices where VDTs and other office machines were in use.

*Noise emission of office machines*

VDTs have brought a number of additional machines into the office: printers, plotters, calculators and others. In Table 31 the noise emissions of some modern office machines are presented.

Table 31 **Approximate noise emissions of office machines.**
*Measurement position about at the head level of the operators.*

| Machine | Noise emissions |
| --- | --- |
| Matrix and daisy wheel printer | |
|   Basic noise | 73–75 dB(A) |
|   Peak levels | 80–82 dB(A) |
| Matrix printer with hood | |
|   Peak levels | 61–62 dB(A) |
| Inkjet printer | |
|   Basic noise | 57–59 dB(A) |
|   Peak levels | 60–62 dB(A) |
| Laser printer | No measurable noise |
| Cooling fans of VDTs | 30–60 dB(A) |
| Old typewriter | ~70 dB(A) |
| Modern electronic typewriter | ~60 dB(A) |
| Two electronic typewriters face to face | 68–73 dB(A) |
| Copying machine | 55–70 dB(A) |

*Matrix and daisy wheel printers are a torture!*

It is obvious that the matrix and daisy wheel printers produced by far the highest noise emissions. Their peak levels, which are highly repetitive, interfere strongly with speech communication and are very annoying, particularly to those who do not benefit from their use. There are some possibilities of reducing this noise: hoods deaden the noise emission by about 20 dB(A); but unfortunately, many employees dislike having to open and shut the hood each time, except when a great amount has to be printed for a long period of time. Where several printers are being used, one should consider locating them in a separate room. The panels which are used to wall off units in large space offices reduce noise emissions by only a few decibels, and they do not damp noise sources exceeding 70 dB.

Inkjet and laser printers are noiseless machines and are greatly appreciated except by those who have to pay the bill!

*The VDT itself is noiseless, but not the cooling fans*

The VDT by itself is nearly silent but many makes have built-in electric fans to cool the unit from the heat produced by the CRT. As long as the noise of these electric fans is below 40 dB(A) it will hardly be noticed, but when it exceeds 50 dB(A) it will become extremely irritating. Personal computers, located in a quiet enclosed space, frequently create this noise problem.

Modern electronic typewriters and keyboards attached to VDTs do not produce excessive noise and are seldom a source of complaint. Most modern copying machines, too, have rather low noise emissions and give no cause to complain, either.

*Conclusion*

To sum up it can be said that computerized offices belong to the category of rather noisy workrooms, but if care is taken to reduce the noise emission of matrix and daisy wheel printers with efficient sound-absorbing hoods or appropriate locations then the computerized office will be within tolerable noise limits, comparable to the earlier large open-plan offices.

# 9. Occupational Stress, Work Satisfaction and Job Design

## 9.1. Occupational stress

*The original definition of stress by H. Selye*

The term stress was introduced by the Canadian Selye after World War II in the field of medicine. *He defined stress as the reaction of the organism to a threatening situation*, and distinguished between the stressor as the external cause and stress as the reaction of the human body. Selye had discovered that stress was essentially a chain of neuro-endocrine mechanisms, beginning with an excitation in the brain stem, followed by an increased secretion of some hormones, especially *adrenalin* and *noradrenalin*. These are referred to as 'performance hormones', since they keep the whole organism in a state of heightened alertness. These performance hormones, also called *catecholamines*, can be determined in the urine, and this is still a possible way of determining stress.

*Physiological reactions*

An increase in activity of this neuro-endocrine alerting system causes the following main physiological stress reactions:

Rise in heart rate
Raised blood pressure
More sugar released by the liver
Increased metabolism

*Health troubles*

These physiological reactions reflect an intensified readiness to defend life, including fighting, fleeing or other physical achievements. But Selye also observed that this emotional state, resulting from the feeling of being threatened, was responsible for the adverse effects of stress. In fact, long-lasting or recurrent stress situations can be detrimental to health by inducing functional troubles, particularly in the gastro-intestinal or in the cardio-vascular systems. These effects are psychosomatic disturbances which, in the long run, can turn into organic illnesses. The most common forms of stress diseases are gastro-intestinal disorders, which can lead to gastric or duodenal ulcers. Selye explained the adverse stress effects on health as a maladaptation of the organism to stress.

*Is stress always harmful?*

It is obvious that stress is part and parcel of our life; it is a necessary condition for all living creatures to react to threatening situations in an appropriate way. A life without stressors and stress would not only be unnatural, but also boring. Stress cannot be divorced from life, just as birth, death, food and love are inseparable.

Paracelsus, a medical doctor of the early sixteenth century, said that *the dose determines whether a compound is toxic or not* (dosis sola facit venenum). The same holds true for stress: the amount determines whether stress will have adverse effects on health or whether it will increase human ability to cope with life. Where the borders between normal physiological and pathological stress are to be drawn is still an open question. Only one thing is certain: this border varies from one individual to another. One person can bear a great amount of stress all his life; another suffers immensely and will sooner or later be overwhelmed by it.

The more the term 'stress' was used, the more it became a myth. Eventually the word was used for nearly every kind of pressure on people. In the last two decades, however, psychologists and social scientists have done detailed research on the phenomenon of stress and formed a much clearer conception of it, particularly with respect to occupational stress.

*Occupational stress*

*The emotional state (or mood) which results from a discrepancy between the level of demand and the person's ability to cope defines occupational stress; it is thus a subjective phenomenon and exists in people's recognition of their inability to cope with the demands of the work situation.*

A stressful situation is a negative emotional experience which can be associated with unpleasant feelings of anxiety, tension, depression, anger, fatigue, lack of vigour and confusion. These feelings characterize the mood, often studied by specially designed questionnaires (Profile of Mood States: POMS).

*Stressors in the work environment*

Surveys as well as theoretical considerations suggest that the following conditions may become stressors in work environments:

1. *Job control* is the worker's participation in determining the work routine, including control over temporal aspects and supervising work processes. Several studies suggest that lack of control may produce emotional and physiological strain.

2. *Social support* means assistance through supervisors and peers. Social support seems to reduce adverse effects of stress. On the other hand, a lack of social support increases the load of stressors.
3. *Job distress or dissatisfaction* is mainly related to job content and work load. It is the perceived stress in job and career.
4. *Task and performance demands* are characterized by the workload, including demands upon attention. Deadlines may be a major stressor, too.
5. *Job security* today refers mainly to the threat of unemployment. Nowadays many office workers worry about being made redundant. Important is the recognition of the availability of similar or alternative employment and of future needs for their professional skills.
6. *Responsibility* for the lives and the well-being of other people may be a heavy mental burden. It seems that jobs with great responsibility are associated with an increased proneness to peptic ulcers and high blood pressure. Responsibility in itself is perhaps not the key stressor. The crucial question is rather whether the amount of responsibility exceeds one's resources.
7. *Physical environmental problems* include noise, poor lighting, indoor climate or small, enclosed offices.
8. *Complexity* is defined as the number of different demands involved in a job. Repetitive and monotonous work is often characterized by a lack of complexity, which seems to be an important predictor of job dissatisfaction. On the other hand, too much complexity can arouse feelings of incompetence and lead to emotional strain.

Any individual may experience a number of other stressors; this list could be easily extended. But the 8 stressors mentioned above are certainly those which are often taken into account by social scientists when preparing their questionnaires to evaluate people's experience with occupational stress.

*Person–Environment Fit*

Research on occupational stressors has come up with the concept of the *Person–Environment Fit*. *The basic assumption is that the degree of fit between the characteristics of a person and the environment can determine the well-being and performance of workers.* Environment is used here in its broadest sense and includes the social as well as the physical environment. Some authors distinguish the fit between the person's needs and their satisfaction through the job environment, others refer to the fit between the demands of the job environment and the relevant worker's ability to meet those demands. In fact, the discussions on the Person–Environment Fit are very similar to those mentioned above about occupational stressors.

Cox (39) writes:

*Measurement of stressors and stress*

Stress, as an individual psychological state, is to do with the way the person sees and then experiences the (work) environment. Because of the nature of the beast, there can be no direct physiological measures of stress. The measurement of stress at work must focus on the individual's psychological state. A first step is thus to ask the person about their emotional experiences or mood in relation to the situation at work. This means using state-dependent subjective data.

Nowadays all field studies on occupational stress are based on extensive questionnaire surveys on working conditions, the worker's health and well-being, potential stressors, job satisfaction and moods. Many authors use scales that have become standardized and widely used instruments for which normative data are available.

Several Swedish scientists also work with physiological parameters of stress, i.e., excretion of catecholamines in the urine, heart rate and blood pressure. Physiological measurements are considered correlates to the questionnaire survey procedures evaluating subjective experience.

## 9.2. Job satisfaction versus boredom

It is undisputed today that job satisfaction is one of the main conditions for fulfilment and pleasure in one's work and a good quality of life in general.

*Boredom*

Job satisfaction depends on all the earlier mentioned stress factors or stressors. One of them, however, seems to be of particular importance: *boredom*.

A monotonous environment is one that is lacking in stimuli, and the individual's reaction to monotony is called boredom. This is a complex psychological state, characterized by symptoms of reduced activity of the higher nervous functions, together with feelings of weariness, lethargy and diminished alertness. Furthermore, symptoms of growing irritability and ill humour may occur, due to the necessity to put up with the monotonous environment. Whatever the reactions may be, boredom is incompatible with job satisfaction and vice versa.

Monotony with the risk of boredom is a phenomenon well-known in modern life as well as in industry and commerce. To drive a car on the motorway is lacking in stimulation and may generate boredom and even sleepiness. An operator at a control desk is exposed to boredom when the signals, requiring a response, occur infrequently. Another example of a monotonous job is being in charge of a stamping press and having to carry out exactly the same operation 10–30 times per minute, for hours, days and years on end. Occupations such as this are *repetitive* as well as *monotonous* and *boring*.

*External causes of boredom*

Experience has shown that the following conditions give rise to boredom:

1. Prolonged repetitive work that is easy to perform, yet which does not allow the operator to think about other things at all; some data-entry tasks at VDTs, like reading numbers from coupons and typing them into the computer, belong to this category of work.
2. Prolonged, monotonous supervisory work, which calls for continuous attention.

It is a fact that certain conditions enhance proneness to boredom, such as a very brief cycle of operations, short training (which means low demand of skill) and little opportunity for bodily movements.

*Personal factors enhancing boredom*

Personal factors have a considerable impact on the incidence of boredom, or, put another way, *on the ability to withstand boredom*. Proneness to boredom is higher for:

People in a state of fatigue.
Not-adapted night workers.
People with low motivation and little interest.
People with a high level of education, knowledge and ability.
Keen people, who are eager for a demanding job.
Extrovert subjects seem to be particularly susceptible to monotonous work.

Conversely, the following are very resistant to boredom:

People who are fresh and alert.
People who are still learning (i.e., a learning driver has no time to be bored).
People who are content with the job because it suits their abilities.

*Taylorism and boredom*

For several years past, more and more critical objections have been raised on the part of social scientists to the Tayloristic principle of splitting a job into a large number of identical tasks which are repeated indefinitely. Working places organized on this principle are characterized by four cycles per piece and few demands on the operative. The result of such a fragmentation of the job is that individual freedom of action is severely curtailed, mental and physical abilities lie fallow and the potential of the worker is wasted.

Several studies (mentioned in 70) have shown that in practice job satisfaction is lower where monotonous, repetitive work is concerned, than with jobs that allow a greater freedom of action and demand more participation as well as more responsibility.

*Some escape into day-dreams and like it*

But the same studies also revealed that there are always individuals who rate their repetitive and monotonous work "interesting"! This confirms the experience of many factories that a certain proportion of the workers enjoy their monotonous, repetitive jobs and do not want one that is more varied or more challenging. It seems that some individuals are able to escape with their thoughts into a world of day-dreams and that they appreciate working conditions which permit them to do so. On the other hand, personnel managers report that it is becoming increasingly difficult to find workers to do the monotonous and repetitive jobs.

Ulich *et al.* (200) made the following comment on this controversial situation:

The contradictory results may perhaps be attributed to differences of attitude that really exist in practice. However, that would mean that for an indefinable proportion of workers, male or female, working on a production line must actually be more relaxing than free assembly, since it allows them to express their personalities better by conversation, day-dreaming and so on. For another, presumably larger proportion of the workers, however, things are quite different. For them continuous work on a production line seems meaningless, and provides them with no opportunities to develop their personalities by exercising their brain power at their work.

*Boredom and adrenalin*

An interesting contribution to a better understanding of the different aspects of monotonous work was made by several Swedish studies like those by Levi (131) and Frankenhäuser (58). They analysed catecholamine excretion in the urine and found that the most diverse physical and emotional stress situations led to a measurable increase in the adrenalin excreted in the urine, which was interpreted as a mobilization of the performance reserves of the body. One study by Frankenhäuser *et al.* (57) is mentioned here because it is relevant to the problem of boredom. Their experiment on mental under- and overload yielded the following results:

*Overload*, created by a long-lasting serial reaction time test, produced an increased flow of adrenalin (about 9.5 ng/min).
*Moderate load*, in the form of reading a newspaper, gave only a small increase in adrenalin excretion (about 4 ng/min).
*Underload*, as a consequence of a uniform, repetitive operation, also produced a higher flow of adrenalin, amounting to about 5.7 ng/min, and so falls between the levels of 'overload' and 'moderate load'.

Frankenhäuser concluded from this:

> The results show that adrenalin production is increased not only when acting under pressure, against the clock, and with a high inflow of information, but also in conditions that are monotonous, and lacking in stimulation. This shows that the physiological reaction is produced by the mental and emotional stress, rather than by the physical effort as such.

*Sawmill work and adrenalin*

A field study by Johansson et al. (98) also produced interesting results: a group of sawmill workers, whose work was repetitive and at the same time responsible, secreted much more adrenalin than other groups of workers. They also exhibited a higher incidence of psychosomatic ailments and more absenteeism. The authors concluded that the combination of monotonous, repetitive work with a higher level of mental stress called for a continuous mobilization of biochemical reserves which, in the long term, adversely affected the general health of the workers.

## 9.3. Alleged stress among VDT operators

*General experience*

Anecdotal reports as well as general experience indicate that the introduction of VDTs in offices also created some psychological problems. In some cases the new technology imposed an increased performance and therefore a greater workload. To illustrate this, the example of payment transfers in a bank is quoted here: without a computer, about 30 payment transfers were settled per employee per working day, but with the aid of the VDT the same employee handled about 300 transfers per day. On the other hand, some new VDT jobs became more and more repetitive and monotonous, especially the data-entry jobs.

*Some worried*

Some employees worried; they were afraid of new technologies, automation and unemployment. This rather complex and only vaguely recognizable situation sometimes gave rise to a general negative attitude towards the new VDT job.

*Some had fun*

However, a contrary reaction was also often observed. In fact, some employees were proud of being included in the new information technology and looked forward to interacting with a computer. Those jobs that required creative participation from the operators were rated interesting. Many managers observed that clerical employees showed some resistance to word processing procedures in the beginning, but after a few weeks clearly preferred the new secretarial work with VDTs to their former job conditions.

In general, it seems that many psychological problems were more acute when VDTs were first introduced and that they are becoming less so as time goes by.

*The Swedish study on stress at VDTs*

One of the first surveys focusing on stress and job satisfaction at VDT workplaces was conducted in 1977 by Johansson and Aronsson (99) on 95 employees of a large insurance company. The extent of daily VDT work ranged from none to more than 75% of the working hours. One group of employees fed data from documents into the computer; others used the VDT for customer contacts; and a third group of subjects had rather complex tasks requiring long experience, namely the settlement of claims, but also some investigative work. The study was conducted in two stages: in the first part a questionnaire was used to obtain indications about attitudes, job satisfaction and experienced stressors. On the basis of the questionnaire data two subgroups were selected for the second part: one group of 12 subjects, a data-entry unit, spent more than 50% of the working time at the VDT. They did routine work, with a low mental load but rather high speed. The other selected group of 11 subjects used the computer not more than 10% of their working time; these employees were typists and secretaries with fairly flexible and varied work, including social interaction. This group served as the control.

*Positive attitudes*

The questionnaire survey of the first part of the study revealed a positive attitude to the job for the whole group and 80% reported personal satisfaction. A majority of 65%, though, thought that the employees had insufficient influence on the introduction of new computer techniques. The attitudes toward computerization revealed a certain anxiety; about one-third feared that continued computerization might render their special know-how superfluous.

*Work load increased with computerization*

On the other hand, only very few considered their own work excessively monotonous or unqualified, but many thought that they had too much to do. Mental strain, the need to concentrate and the amount of routine work were all reported to have increased. However, most respondents claimed that they had a better overall view of an insurance case.

*Response time should not exceed five seconds*

The response time of the computer was frequently criticized. *As many as 63% of the subjects were of the opinion that the computer response time should not exceed 5 s.*

A group of 34 questions dealt with stress factors: high workload, computer breakdowns and uncertainty as to the duration of breakdown got the highest ratings. Conversely, the highest mean ratings for contributing to job satisfaction were obtained for job security, flexible working hours and independence at the job.

*Physiological stress reactions*

The second part of the study focused on physiological reactions of the two selected subgroups, including measurements of adrenalin and noradrenalin (catecholamines), heart rate and blood pressure. At the same time the subjects made quantitative assessments of arousal and mood on a scale ranging from "not at all" to "maximally".

The results indicated that the data-entry group had slightly higher catecholamine levels than the control group, but both disclosed a moderate excretion of stress hormones. The heart rate was slightly higher for the control group, but the difference was not significant.

*Mental fatigue*

The questionnaire data revealed that the VDT group was to a much greater extent affected by mental strain than the control group. These results are shown in Figure 94.

The self-ratings of rush, irritation and fatigue resulted in slightly higher figures for the VDT group compared to the control group.

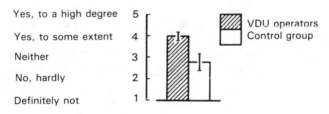

**Figure 94 Mental fatigue after the end of work for the VDT operators and the control group.**
*"Do you feel worn out after the day's work?" Data-entry VDT group = 12 subjects; control group = 11 subjects. According to Johansson and Aronsson (99).*

*Computer breakdown and stress*

The most striking result of this study was observed during a temporary breakdown of the computer system. Reliable data were obtained from six individuals during this four hour breakdown. The results are compiled in Figure 95.

The presented data show that adrenalin, blood pressure and heart rate were raised during the breakdown, compared to the usual condition. The differences were significant for adrenalin and diastolic blood pressure. At the same time the subjects felt more irritated, tired, rushed and bored. The authors *concluded that computer breakdowns are an important cause of mental strain for persons with extensive VDT work.* In fact, an interruption meant that the VDT operators were condemned to idleness while their own work was piling up, which presumably increased the next day's work load. The authors believe that stress and strain at VDTs may be partly counteracted at the technological and the organizational level: by reducing the duration and frequency of breakdowns, by

# Ergonomics in Computerized Offices

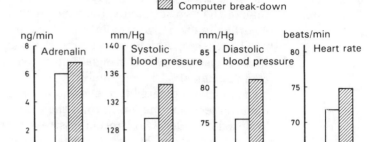

Figure 95  Mean excretion of adrenalin, systolic and diastolic blood pressure and heart rate during a 4-hour computer breakdown and during regular work at the same time of day.
n = 6 subjects. According to Johansson and Aronsson (99).

shortening the response times in the system and by redistributing unavoidable but monotonous data-entry work.

*The study by Elias and Cail*

The French study by Elias and Cail (51) has already been referred to in Figure 4, comparing the frequency of scanning and the look duration at a data-entry and a conversational task. In the course of this study the authors also investigated psychosocial aspects related to job satisfaction. One group of 89 female subjects was engaged in a data acquisition job in a bank, which was characterized by monotonous and repetitive work. The other group of 81 female subjects had a conversational task in a publishing and a pharmaceutical company. The questionnaire included four categories of questions relative to visual strain, bodily pain, neuropsychological disturbances and job satisfaction.

The incidence of reported job dissatisfaction is presented in Figure 96.

The main complaints about the work were different in both groups; they are reported in Table 32.

The major functional health troubles were related to gastro-intestinal symptoms, anxiety, irritation and sleep disturbances. All these items were significantly more frequent among the data acquisition group than in the conversational group.

*Correlations not convincing*

A correlation analysis revealed in both groups a significant relationship between monotony and job dissatisfaction ($r = 0.588$). In the data acquisition group, however, there was no significant correlation between the individual stressors and mood symptoms (anxiety, irritability or state of depression) whereas in the conversational group significant correlations were found between:

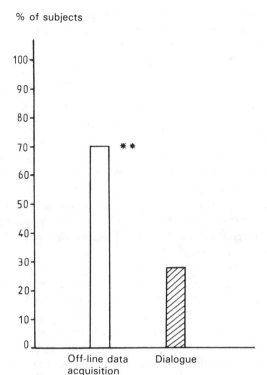

**Figure 96 Job dissatisfaction reported by VDT operators.**
*White column=data acquisition operators; striped column=conversational operators.* **=significant at p < 0.01. According to Elias and Cail (51).*

**Table 32 Main complaints about the work in a data acquisition task and in a conversational job.**
*According to Elias and Cail (51).*

|  | % of subject |
|---|---|
| *Data acquisition task (n=89)* | |
| Time pressure | 80 |
| Monotony | 80 |
| Difficult contacts with colleagues | 30 |
| Job satisfaction | 30 |
| *Conversational task (n=81)* | |
| Work interruptions due to breakdowns of the computer system | 90 |
| Monotony | 65 |
| Insufficient possibility of controlling and correcting errors | 30 |
| Job satisfaction | 70 |

Work interruptions and psychosomatic troubles ($r=0.28$)
Monotony and anxiety ($r=0.23$)
Monotony and irritability ($r=0.42$).

The authors point out that the operators with the data acquisition job often handle data without knowing what they mean; such a highly fragmented task with poor cognitive content does not stimulate interest and does not make use of the operator's abilities. These characteristics may explain the higher frequency of psychosomatic disorders and inadequate sleep patterns in the data acquisition group.

*The NIOSH studies*

Smith et al. (183, 184) carried out two field studies to evaluate job stress among VDT operators and control subjects. A questionnaire survey was used to gather information about job demands, job stressors, psychosocial stress, psychological mood and health complaints. The subjects of the first study were from three different offices in the San Francisco area. The questionnaires were filled out at home, and the response rates were rather low (between 43 and 73%). The responses of approximately 250 VDT operators and 150 control subjects were analysed.

*VDT operators and control subjects reported an increase of psychosocial stressors*

It was found that the VDT operators from all offices as well as the control subjects were subjected to a large number of psychosocial stressors, which indicated more stress exposure than one would have expected from established normative values. In general, the VDT operators reported more psychosocial stress than the control subjects, but this tendency was more pronounced at office 1 than at the other two sites.

Health complaints and illnesses were collected with the aid of a self-report checklist. None of the states of disease showed a significant difference between the VDT operators and their controls from either of the three sites, but some significant differences became obvious when looking at the sites individually. Moreover, the first NIOSH study had a number of shortcomings:

*Critical objections*

The jobs were not described and differing tasks could not be compared.

Low response rates raise the question whether the groups were at all representative.

At the time of the study labour negotiations were in progress, which could have caused additional strain.

*The second NIOSH study*

Smith et al. (184) soon afterwards conducted a second field study with the same goal as the first. The investigation was carried out at five workplaces, namely four newspaper publishers and one insurance company. The VDT operators did various jobs including data-entry and retrieval, word processing, writing, editing and telephone sales. The control subjects were engaged in similar jobs but without VDTs. The respondents to the questionnaire survey were divided into three groups: *professionals* (reporters, editors) using VDTs and endowed with a great amount of control over work and

*Low response rates*

activities; *clerical VDT operators* with data-entry and data-retrieval jobs, closely supervised, with little control over the work procedure; and *control subjects* with jobs identical to those of the clerical VDT operators, but without computer. A total of 250 VDT operators and 150 control subjects took part in that survey. The response rates were rather low: 50% of the VDT operators and 38% of the control subjects. The questionnaires were filled out at home. They resembled those of the first study, including standardized scales for job stress and psychological mood. The results can be summarized as follows:

1. The clerical VDT group reported less peer cohesion and job autonomy and more work pressure than the professionals using VDTs and the control subjects.
2. The clerical VDT operators reported higher workload, more boredom, greater workload dissatisfaction, greater job future ambiguity and lower self-esteem than the two other groups.
3. The mood scales, however, revealed no significant differences among the three groups, except for "fatigue", which was rated slightly higher by the clerical VDT operators.
4. The clerical VDT operators reported more health complaints that the two other groups. Typical stress-related health troubles, like irritability, nervousness and stomachache were more frequent among clerical VDT operators in comparison to the control subjects, but not when compared to the professionals.

*Job content is a major cause of increased stress*

The authors believe that the job content of the clerical VDT operators may be an important factor in increased occupational stress and health complaints. Indeed, the clerical VDT operators held jobs involving rigid work procedures, constant pressure for performance, little operator control over the job and little satisfaction from the end product of their work. For the professionals the VDT was a tool that enhanced the end product, whereas for the clerical VDT operators the terminal was part of a new technology that took more and more meaning out of their work. In fact, the professionals using VDTs reported the lowest stress rates, the clerical VDT operators the highest, and the control subjects figured in between. According to the authors this suggests that the use of VDTs is not the only factor contributing to operator stress and health complaints, but that job content, too, plays an important role in this matter.

*Critical objections*

The authors themselves make the following restrictions:

This evaluation may have limited generalizability, since the study sites were not selected at random, but rather were known sources of union complaints about health problems. Moreover, participants were not selected randomly, but were volunteers, and difficult labour negotiations were underway at the same time of the data collection... One explanation for this heightened stress level for both the VDT operators and control subjects is that it may have been due to the strained employee/management relations.

The second NIOSH study, too, had low response rates. Again, this raises the possibility that the non-respondents may be significantly different from the respondents in key ways.

As mentioned in Section 9.1, occupational stress is defined by social scientists as the emotional state resulting from a discrepancy between the level of demand and the person's ability to cope with these demands. On the basis of this conception of stress one should expect that excessive stressors or demands cause corresponding changes of mood in the sense of heightened anxiety, depression, anger, nervousness and other emotional reactions. But such relationships were not found in the NIOSH studies and it is therefore doubtful whether the reported increased stress levels (which, in fact, were psychosocial stressors) of VDT operators can really be considered increased stress in the sense of the above mentioned definition of the term.

*The study by Dainoff et al.*

A survey on 121 office workers using VDTs with open-end interviews was carried out in the Ohio area by Dainoff et al. (41). Their aim was to obtain some direct indications on the attitudes of the VDT operators towards their work and towards office automation. A smaller subgroup was examined more closely, in particular with a questionnaire on mood and physical symptoms. No control group was involved and a comparison with non-VDT groups is missing. The VDT operators were engaged in word processing, financial operations, record retrieval and data-entry operations. The lack of a control group makes any general conclusion difficult. Nevertheless, the comments of the VDT operators on their work were interesting and can be summarized as follows:

*Attitudes towards VDT work mainly positive*

56% liked their jobs, by and large.
81% found the computer technology to be efficient.
36% had fun working with VDTs.
43% complained about the poorly arranged workstation.
49% deplored work interruptions caused by computer breakdowns.
6% expressed hostility toward the computer or indicated fears about employment.

*Visual fatigue independent of attitudes*

In addition to these findings, the study revealed a high incidence of eye fatigue symptoms as well as complaints regarding glare and lighting. However, the patterning of these complaints was independent of job pressure and hostility towards office computerization and the authors concluded that the reported visual symptoms could not be explained by non-visual aspects of the job, such as pressure and employee hostility towards the computer.

*The Wisconsin NIOSH study by Sauter et al.*

A survey, carried out by Sauter et al. (177, 178) and fully supported by NIOSH, focused on job attitude, affective and somatic manifestations of stress among VDT office workers. The study consisted of an extensive questionnaire survey of working conditions, job stressors, mood states, job dissatisfaction and health symptoms. Scales were used to assess the greater part of the stressors and the mood states. The methodology was similar to the one employed in the previous, above mentioned NIOSH studies. The VDT group consisted of 248 subjects, the control group of 85. The VDT operators worked at the terminal more than four hours per day. The control group had almost identical tasks but had to carry them out without VDTs. More than 90% of the solicited employees participated in the survey. Their tasks included above all data- and text entry, file maintenance, word processing and general clerical secretarial work. At the time when this study was carried out, VDT use was not an issue in labour negotiations for these employees in the Wisconsin area.

The analysis of the questionnaires yielded the following results:

*No indication of stress in either group...*

1. *None of the well-being indices, related to job stressors and mood states, disclosed a strong indication of increased strain for the VDT group.*
2. There were no marked group differences in terms of job or work load dissatisfaction or boredom. But in both groups the indices of boredom were higher than for 23 occupational samples determined by Caplan et al. (31).
3. The VDT users had a higher incidence of unfavourable working conditions; they reported greater under-utilization of skills, lower supervisory and staff support, and lower job autonomy. However, the VDT group reported less role conflict and more control over working processes.

*...but VDT users reported unfavourable working conditions*

4. VDT users rated their working place environment less pleasant and their chairs less comfortable, and they were more bothered by smell, dust and constrictions than the control subjects.
5. For VDT users the chairs and the workstation setting in particular were important predictors of musculo-skeletal disturbances, just as ambient lighting was of visual discomfort. But the extent of screen viewing as such was

not predictive of either visual or musculo-skeletal discomfort.

The authors summarize their findings as follows: *"Other than a tenuous indication of increased eye strain and reduced psychological disturbances among VDT users, the two groups were largely undifferentiated on job-attitudinal, affective, and somatic manifestations of stress".*

*The Starr studies on job satisfaction showed...*

The two studies by Starr et al. (192, 193) on the effects of VDTs on telephone operators and on office employees have already been discussed in Section 6.1. Here only their results related to job security and job satisfaction shall be mentioned The telephone operators using VDTs and their control subjects were asked several questions related to job security. Most subjects in both groups felt that they were not going to be laid off and that the classification of their job would not be changed. About 25% of the subjects in both groups imagined that their job would become less important in the course of automation. About equal proportions said they would choose to remain in or leave their present job within six months. Job satisfaction was studied with a special "Job Descriptive Index" and disclosed significantly greater satisfaction among VDT operators than among the control subjects. Supervision was regarded with approximately equal satisfaction by both groups. A comparison of the item "age" yielded merely one significant difference: the VDT group retained their higher satisfaction on the checklist. Another questionnaire — the "Job Dimension Checklist" — confirmed the above mentioned results.

*...greater satisfaction among VDT operators*

The study on office employees (193) yielded similar results: the VDT users were more satisfied with their work and their chances for promotion, whereas the control subjects were more satisfied with people they met in the job. The authors concluded that replacing paper documents by VDTs need not adversely affect the morale of office workers.

*Criticism of the US National Academy of Sciences*

The panel of experts of the US National Academy of Sciences (153) analysed the first five of the above described studies (41, 51, 99, 183, 184); those of Sauter (177) and Starr (192, 193) were not yet published at that time. The experts express sharp criticism on the methods of the surveys. In particular, they denounce the following.

1. The lack of a theory, which causes an arbitrary choice of the stressors to be studied.
2. Cross-sectional instead of longitudinal surveys.
3. The inadequate sampling of VDT and control groups; most samples are not representative in their view and appear to be ill-suited for drawing references for the broader population.
4. The low response rates, which raise the possibility that the

respondents are a non-representative selection of negative or positive attitudes.
5. Statistical shortcomings of correlations between stressors and stress symptoms.
6. A lack of distinction between different types of VDT jobs; sometimes employees of diverse sections of the population are sampled and treated as if they were similar to VDT users.
7. Most stressors are not inherent in VDT technology and software, but depend to a large degree on the type of work and its organization.

*Literature on psychosocial stressors is inconclusive*

The experts conclude as follows:

There is a great deal of freedom to make work at a VDT as pleasurable or as painful as work at a desk with a typewriter. *If there are health risks inherent in VDT work that derive from psychosocial stressors, there is no compelling evidence in the literature (published before 1983)..... the VDT literature on psychosocial stressors is, with a few exceptions, inconclusive because the designs of the studies have not allowed conclusiveness.*

This conclusion is certainly too severe! The above described studies disclosed a number of interesting results which deserve attention and careful consideration. One should remember the fact that shortcomings in the methodology do not prove that the recorded results are wrong, they merely refer to the limitations of interpretation and general conclusions.

*Arguments in defence of scientists*

Field studies are seldom perfect and criticism on insufficient scientific accuracy is always possible. When planning and conducting a field study one is usually confronted with many practical difficulties and every scientist knows from experience that such research always involves compromises. A great problem is often the use of rigorous sampling methods. Many managers or employers object to questionnaire surveys because of the risk of social disturbances. Thus the researcher often has no other possibility than accepting the population available for study.

*Conclusions*

Table 33 summarizes the main results obtained from the eight field studies focusing on occupational stress and job satisfaction in VDT jobs. In this summary the term 'stress' is understood as the emotional mood due to the discrepancy between stressors and the person's ability to cope with them.

In spite of the above mentioned methodological deficiencies it is possible to draw some tentative conclusions:

1. *Generally speaking, clerical VDT operators do not show symptoms of excessive stress.*
2. *There is one important exception, though: some VDT*

Table 33 Summary of the main results, obtained in eight field studies, focusing on psychosocial and environmental stressors, job satisfaction and occupational stress at VDT workstations.

| Citation | Main results |
| --- | --- |
| Johansson and Aronsson (99) | The whole VDT group revealed a positive attitude and the majority reported personal satisfaction with their work at the VDT. Excessive work loads and breakdowns of the computer system were major sources of complaints. A subgroup of VDT operators showed clear stress reactions with increased catecholamine excretion and heart rate during a long-lasting computer breakdown. |
| Elias and Cail (51) | Main complaints of a VDT group engaged in data acquisition time pressure, monotony and job dissatisfaction. Main complaints of a conversational VDT group: computer breakdowns and monotony. Data acquisition revealed more stress-related psychosomatic troubles and adverse mood changes. High correlation between 'monotony' and 'job dissatisfaction'. Other correlations not conclusive. |
| Smith et al. (183, 184) | In a first study VDT and control subjects reported an increase in psychosocial stressors. In the second study, a clerical VDT group reported more psychosocial stressors and job dissatisfaction than professionals using VDTs and control subjects; but there was no difference in the mood scales. |
| Dainoff et al. (41) | Open-ended interviews revealed mainly positive attitudes towards VDT work. |
| Sauter et al. (177, 178) | None of the well-being indices related to job stressors and mood states disclosed a strong indication of increased stress among the VDT workers, although they reported more unfavourable working conditions than control subjects. The two groups showed no difference as to job satisfaction, affective and somatic manifestations of stress. |
| Starr et al. (192, 193) | A first study showed that telephone operators using VDTs were more satisfied with their work than control subjects. A second study on office employees, using VDTs, disclosed the same tendencies. |

*operators, engaged in very fragmented, repetitive and monotonous jobs, such as data-entry or data acquisition, experience on average stronger psychosocial stressors, report low job satisfaction and indicate a higher frequency of mood changes for the worse, as well as gastro-intestinal or other psychosomatic troubles.*

3. *The fact that other studies, including repetitive VDT jobs, did not reveal more psychosocial stressors or symptoms of stress than control groups, leads to the conclusion that it is not the work with the VDT as such but the poor work structure of some specific repetitive jobs which is responsible for the observed adverse effects.*
4. *Apart from repetitive and monotonous jobs, VDT operators are, on the whole, satisfied with their work and consider a VDT an efficient tool.*
5. *Breakdowns of the computer system seem to be, without exception, a very annoying condition, which in one case (99)*

*even produced clear stress symptoms with increased heart rate and catecholamine excretion.*

These conclusions do not confirm the views of many authors who believe that stress is prevalent among clerical workers and VDT users (36). One reason for this disagreement is certainly that many authors have different or sometimes unclear conceptions of the phenomenon of stress; they often do not distinguish between stressors and stress. Such confusion becomes particularly evident when eye strain or physical discomfort in the neck—shoulder area are regarded as symptoms of stress.

## 9.4. Job design

There are good reasons for believing that a job that takes account of a person's abilities and inclinations will be carried out with interest, satisfaction and good motivation. On the other hand, it is obvious that an undemanding job, which does not develop the potential of the worker, will turn out to be boring and lacking in stimulation, and at the other extreme, a job that requires more from the worker than he is capable of will be too much for him/her. Hence the requirement that *work should be so planned that it is equal to the abilities of the operative, without asking too little or too much of him/her.*

*Work load neither too low nor too high*

Many social scientists and occupational psychologists agree that *work efficiency, job satisfaction and work motivation are at their highest when in the range between being under- and overdemanding.*

For office jobs it is the complexity that determines whether a task represents a mental under- or overload for an employee. The relationship between job demands or complexity and job satisfaction and work efficiency is illustrated in Figure 97.

The bell-shaped curve of this figure has only a very general meaning: a moderate and therefore optimal degree of com-

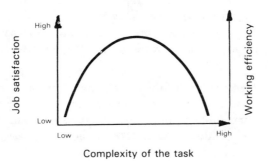

**Figure 97** Conjectured relationship between the level of complexity of a job and working efficiency or job satisfaction.

plexity might be too low for some well motivated subjects, but too high for less ambitious and less skilled persons. In other words, the bell-shaped curve can shift considerably depending on the character and personality of the person involved.

*Main effects of Taylorized jobs*

It has been repeatedly pointed out that fragmented, repetitive, meaningless and monotonous work is usually associated with boredom. Moreover, Frankenhäuser *et al.* (57) found that underload, too, produces a higher flow of adrenalin, indicating a stress-like reaction. It is interesting to note all the adverse effects that are attributed to monotonous work by the different scientific guilds! Table 34 compiles the most important critical objections, culled from different scientific disciplines, to an extreme Tayloristic fragmentation of work into monotonous and repetitive tasks.

The consequence that may result from repetitive work as set out in Table 34 have, in recent years, led to the development of different ways of organizing and restructuring assembly work and similar serial jobs. There is no doubt that some data-entry or data acquisition jobs at VDTs belong to this type of repetitive and meaningless work.

Table 34 **Fragmented, repetitive, meaningless and monotonous work as perceived by different sciences.**

| As seen by: | Probable consequences |
|---|---|
| Medicine | Atrophy of mental and physical powers |
| Work physiology | Boredom; risk of errors and accidents |
| Psychology | Human potentiality not fully exploited |
| Social science | Job dissatisfaction |
| Work science | Increased absenteeism; difficulty in finding personnel to do the job. |

*Aims of restructuring job design*

The main objective of efforts aimed at improving job designs is to give the operator more freedom of action in the following two ways:

1. *Reduction of boredom, with its secondary phenomena of fatigue and job dissatisfaction.*
2. *Making the work more worthwhile by providing a meaningful job, which allows the operator to fully develop his potential.*

Basic to these aims is the assumption, indicated in Table 34, that there will be a reduction of absenteeism, work force turnover and psychosocial stressors, and that the new working conditions will attract more workers. Hence, in the long run, higher productivity should result.

Such desirable new forms of work organization involve various improvements, *ranging from a simple increase of work* variety through different ways of broadening the scope of the job, to job enrichment by giving the worker more information, more responsibility, more participation in decision-making and more control of the work process.

*Increase of variety of work*

An important first step to improve repetitive working conditions is the attempt to increase the variety of work. This is a scheme by which each individual worker is entrusted with different jobs at different working places, which he carries out in rotation. Such attempts have been made in industry, for instance the rotation of workers round a work bench with frequent changes of assembling operations. The best solutions were judged to be those which at the same time introduced a certain group autonomy, allowing the workers to control the making of their own product.

One point must be emphasized, though: if variety of work merely means moving to and fro between jobs that are equally monotonous or repetitive, the risk of boredom may be slightly reduced, but the desirable matching of the difficulty of the job to the capabilities of the worker is not being achieved. Adding yet another monotonous, repetitive job is not going to lead to job enrichment!

*Broadening and enriching the job*

Several recommendations of social scientists and psychologists attach special importance to those types of work organization which strive to enrich the work by broadening its scope, thus helping to develop the personality and self-realization of an employee. Many examples of such broadening and enrichment of employment in industry can be found in the literature (60, 168, 200), but no systematic attempts are known favouring the enrichment of office jobs.

*Autonomous working groups*

A further step towards more participation by workers was the organization of autonomous working groups. The workers employed for each production unit were organized in a group and the planning and organization of the work as well as the control of the end product were delegated to them. The autonomous working group, therefore, has planning and control functions.

All the efforts to broaden and enrich people's work must be regarded as experiments that are by no means completed, nor can their results as yet be fully evaluated. There are reports on successful attempts, but there are also cases where such projects had to be given up because of resistance on the part of employees, unions or managers. In fact, the search for new ways of restructuring monotonous, repetitive and meaningless work is still going on. So far offices have been excluded from such projects.

The importance of social contacts at the working place can hardly be overestimated. *The opportunity to talk to one's fellow workers is an effective way of avoiding boredom.* Conversely, social isolation brings monotony and enhances the tendency to become bored with one's work.

*And what about VDT jobs?*

As shown above, in industry attempts have been made to avoid the adverse effects of repetitive, monotonous jobs by means of alternative ways of organizing production. Johansson (36) writes:

> Unless counter-measures are taken also in the white-collar sector *there is a risk that the new technology* [of VDTs] *will create highly repetitive tasks which will require little skill, allow little social interaction, and generate the type of negative consequences associated with mechanized mass production. Data-entry work is a case in point.* (Emphasis added)

Many authors agree with this word of warning and point out that some clerical jobs had little content before the era of VDTs, and that VDT jobs are often fragmented and simplified versions of traditional clerical activities. This is certainly true for some very simplified data entry or data acquisition tasks, whereas many other VDT jobs are characterized by a high degree of complexity, judged interesting and challenging by operators.

*Job redesign necessary for data entry or data acquisition*

*Attempts to improve job design are therefore justified mainly for the highly repetitive and monotonous data entry and data acquisition jobs.*

No projects or results of restructuring fragmented and repetitive VDT jobs with the aim of improving job design have been published yet. For that reason claims in connection with job design are mainly based on general considerations of the relationships between working conditions and job satisfaction.

*Broadening VDT tasks through 'mixed activities'*

Broadening data-entry tasks is certainly rather difficult. In principle, one should avoid excessively repetitive work where merely simple perceptual motor skills are used, but not social or cognitive skills. In fact, some banks improved the situation by creating 'mixed jobs', alternating pure data-entry with payment transfers and other more demanding tasks. Furthermore, rest rooms were provided where operators could take their breaks together. Other banks engaged only part-time employees for pure data-entry jobs, which, however, was less successful than 'mixing activities'.

*Control of work process*

Lack of job control seems to be an important social stressor. Job control means participation and increased decision making. Performance feedback is a vital part of the worker's control of the work process. It is recommended that an operator should receive feedback direct via his/her own VDT screen. Such direct performance feedback seems to enhance job satisfaction. If, however this information is forwarded to the supervisor, the employee will experience this as lack of control of his work; the supervisor might use the information to control the worker, thus creating more social tensions.

*VDT work should be meaningful*

The meaning or content of work may be low in some VDT tasks. As work is fragmented, it is also simplified. Thus workers fail to identify with the job and loose interest in the product of their work. If fragmentation cannot be avoided, it is important that employees at least understand their contribution to the end product. They should feel that their contribution is important, which will heighten their satisfaction and self-esteem.

*Social contacts should be granted*

One of the obvious drawbacks of VDT work is the poor opportunity for social contacts, particularly with colleagues. Many VDT jobs lead to an isolation of individual operators, much more than traditional clerical activities. Therefore it is advisable to enhance and encourage social interaction during non-task periods; this is an argument in favour of work breaks with shared facilities nearby.

*A careful introduction policy*

A careful and well-planned introduction of VDTs is an important measure to prevent hostility towards office automation. A good transition policy should include proper information and clear instructions, adapted to the worker's capabilities. It is certainly insufficient to leave the employee alone with a manual that explains how the system works. Classroom teaching, followed by practical application, should be given by a well-qualified expert able to coach the trainees. A good training programme will increase acceptance and reduce psychological fears, since well-trained operators will come to consider themselves an important investment.

*Some restrictions must be emphasized*

Some restrictions must be made, though, when discussing the above mentioned principles of job design for VDT work. It must not be overlooked that the main problems of job design refer only to some repetitive, monotonous and meaningless data-entry or data acquisition jobs. The proportion of these among the total number of VDT activities is not known, but a figure of between 15 and 25% is a reasonable assumption. Another restriction concerns the fact that not every employee dislikes repetitive jobs. Salvendy (176) reports that 10% of the labour force in America does not like work of any type; the remaining workers are evenly split between those who prefer to work in enriched jobs and find them satisfying and productive and those who prefer to do simpler jobs in which they are more satisfied and productive. When designing office jobs, he

*Not every repetitive task is monotonous*

concludes, the workers should be given a choice between either a simpler or a more challenging job. A third restriction suggests itself, considering that not every repetitive task is monotonous. For example the data acquisition tasks of telephonists engaged in directory assistance operations: they have about 30 seconds per customer call, but every call involves a new voice with a different accent and an unforeseen question. This job is certainly repetitive, but not at all monotonous. It is therefore not surprising that Starr *et al.* (192) found greater job satisfaction among telephone operators using VDTs than among control subjects doing the same job without VDTs.

*Working hours at VDTs*

The schedule of working hours and breaks is an important aspect of job design. The results, related to the impact of working hours on the adverse effects of VDT work are controversial: some observed an increase of complaints with increasing working time at a VDT, others could not confirm such a relationship. Some unions and a few scientists claimed that the number of working hours each day spent at VDTs needed to be reduced. A principle objection must be made here: *neither change nor reduction in working time should be envisaged as long as the display, workstation and environment do not fulfil the main ergonomic design recommendations.* It would be nonsense to reduce the working time because of a badly designed workplace. There are good reasons, at present, to believe that the work at ergonomically well-designed VDT workstations is not more strenuous than other office jobs.

*Rest pauses*

Every function of the human body can be seen as a rhythmical balance between energy consumption and energy replacement, between work and rest. This dual process is an integral part of the operation of muscles, the heart and the organism as a whole. *Rest pauses are therefore indispensable as a physiological requirement, if performance and efficiency are to be maintained. They are essential, not only during manual work, but equally during every type of mental work.*

*There are different kinds of rest pauses*

Ergonomic studies have shown that people at work take rest pauses of various kinds and under varying circumstances. Four types can be distinguished:

> *Spontaneous breaks*
> *Disguised pauses*
> *Work-conditioned pauses*
> *Prescribed breaks*

*Spontaneous breaks* are interruptions for rest that the workers take on their own initiative. These breaks are usually not very long, but may be frequent if the job is strenuous.

*Disguised pauses* are periods when the worker occupies himself with some easier secondary activities in order to relax from concentration on the main job. Office jobs offer many opportunities for such disguised pauses, for example, cleaning office machines, tidying the desk, blowing one's nose, or even leaving the working place on the pretext of consulting a workmate or a supervisor. *Such disguised pauses are justified from a physiological point of view, since nobody can do either manual or mental work continuously, without interruption.*

*Work-conditioned pauses* are all those interruptions that arise either from a machine (like computer break-downs) or from the organization of the work (e.g., waiting for orders, documents or customers).

*Prescribed breaks* are breaks in work that are laid down by the management.

*Interrelationships between pauses*

The four types of rest pauses are to some extent interrelated: the introduction of prescribed pauses leads to fewer spontaneous and disguised pauses. The same effect is to be expected from excessive work-conditioned pauses. Studies have shown that spontaneous and prescribed pauses increase considerably during the afternoon, when operators are getting more tired.

In industry it is a general rule that all the different types of pauses should amount to about 15% of the total working time. For some strenuous jobs a ratio of 20–30% is often allowed for.

*Rest pauses and output*

Many investigations of the effect of rest pauses on productivity have been carried out in industry (70). They showed that the introduction of prescribed rest pauses speeds up the work, and this compensates for the time lost during prescribed pauses, as well as leading to fewer disguised and spontaneous pauses.

The hourly output of fatiguing work usually declines towards the end of the morning shift, and even more towards the evening, as the working speed slows down. Various studies showed that if prescribed pauses are introduced, the occurrence of fatigue is postponed and the loss of production through fatigue is less. On the whole, rest pauses tend to increase output rather than decrease it.

One rule applies here: *no well organized and attention-requiring work can be carried out without interruptions; if prescribed rest pauses are not allowed, blue collar and white collar workers will make the necessary interruptions to relax in the form of more spontaneous and disguised pauses.*

For all jobs in offices and in administration the current recommendation is a rest pause of 10–15 minutes in the morning, and often a similar interval during the afternoon. These pauses serve the following purposes:

*To prevent fatigue*
*To allow opportunities for refreshment*
*To provide time for social contacts.*

Summarizing, the following arrangement of rest periods is to be recommended:

*Recommendations: four working periods per day, separated by two rest pauses and a lunch break*

*For office jobs demanding attention and flawless performance it is advisable to divide the daily work into four equal periods: two in the morning shift, separated by one rest pause of 10–15 minutes, and two in the afternoon shift, again separated by a rest pause. A longer pause of about 45 minutes at noon should be fixed between the two shifts, allowing enough time for lunch.*

*Additional short breaks of 3–5 minutes every hour might be indicated when learning a skill or serving an apprenticeship, and for jobs with great mental demands, especially if the work is machine-paced.*

*Most VDT jobs certainly belong to the above-mentioned office jobs. Two rest pauses, one in the morning and one in the afternoon shift, are recommended.*

*Repetitive and attention requiring VDT jobs, such as data entry or data acquisition, should first be redesigned in the sense of broadening the task; and if this is not possible, additional short breaks should be allowed for.*

# 10. Radiation, Electrostatic Fields and Alleged Health Hazards

## 10.1. Electromagnetic radiation emission from VDTs

*Electromagnetic waves*

Electromagnetic radiation is the propagation of energy in wave form; some of its effects are essential for the existence of life on Earth. Electromagnetic radiation can be reflected or absorbed by materials and bodies, whereby useful or injurious effects may occur.

The number of wave peaks per second expresses the frequency of hertz (Hz), which is the most commonly used parameter to characterize electromagnetic radiation. Its range can be visualized as a continuum, called the electromagnetic spectrum, where the radiation is ordered according to frequency, as shown in Figure 98.

Electromagnetic radiation may be divided into two parts, as a function of its biological effects, namely into an ionizing and a non-ionizing area.

Ionizing radiation possesses sufficient energy to disrupt and separate electrons from atoms. In this process ions are

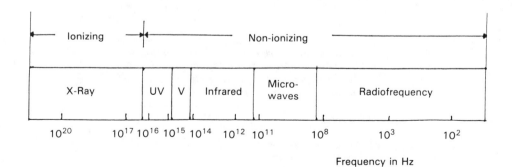

**Figure 98** The spectrum of electromagnetic radiation.
*UV = Ultraviolet; V = Visible light.*

generated, forming the essential component of ionizing radiation. The energy of electromagnetic radiation is proportional to its frequency: when the frequency is reduced, the potentially damaging energy is also diminished.

*Ionizing radiation (or X-rays)*

In the cathode ray tube of VDTs, electrons are generated inside the tube and directed by the electron beam onto the inner surface of the front glass of the tube.

The acceleration of the electrons is due to a high positive voltage on the accelerating anodes inside the tube. The voltage level determines both the degree of electron acceleration and electron energy.

Fortunately, the ionizing radiation, also called X-rays, generated inside the cathode ray tubes, is of relatively low energy. These X-rays are reduced to negligible numbers by correct circuit design and a high lead content in the glass front of the tube.

The biological effects of X-rays have been known for a long time. A continual exposure to them at sufficient dosage can cause cancer, genetic damage and other injuries. The effects tend to be cumulative and irreversible, so that the total absorbed dose is the relevant criterion.

*X-ray emissions from VDTs...*

X-ray emissions from cathode ray tubes are described in terms of the exposure rate in milliroentgens per hour (mR/h). *Assuming a five-day and forty-hour week, official regulations for personnel occupational safety throughout Europe and the United States set a maximum permissible exposure of 2.5 mR/h at a distance of 40 cm from the screen.* The intensity of the background X-ray radiation of natural origin is of the magnitude of 0.01–0.03 mR/h.

Many studies on X-ray emissions of VDTs have been carried out by official agencies as well as by university scientists in various parts of the world. Most of this research work has been analysed by the US Academy of Sciences (153) and, more recently, by Bergqvist (16). For once, all the studies reached the same conclusion: *emissions of X-rays were well below*

*...are far below health standards limits*

Table 35 Measurements of X-ray radiations.
*All values represent the highest reading from any VDT set tested 5 cm from the screen surface, in milliroentgen per hour.*

| Reference | mR/h |
|---|---|
| Weiss and Petersen (207) | <0.5 |
| Terrana et al. (196) | 0.02 |
| Cox (38) | 0.01 |
| Bureau of Radiological Health (25) | Not detectable |
| NIOSH investigation, 1981 (154) | Not detectable |

*accepted occupational and environmental health and safety standard limits.* A few of the results are reported in Table 35.

In some field studies the measured X-ray emissions were slightly higher, because of the presence of higher than average levels of natural ambient radiation.

The US National Academy of Sciences (153) summarized the results of all the studies analysed as follows:

*The conclusions of the US National Academy of Sciences*

*Taken collectively, these studies have examined a wide variety of models and hundreds of terminals. Measurements of emissions from older and newer VDT models have not differed significantly. Measurements have been made both under normal operating conditions and under conditions designed to maximize potential emissions. ... These studies have concluded that the levels of all types of electromagnetic radiation emitted are below existing occupational and environmental health and safety standard limits of exposure ... The level of X-ray radiation emitted by VDTs is far less than the ambient background level of ionizing radiation from natural sources to which the general population is exposed.*

A more recent analysis in 1984 of all the studies on X-ray exposure at VDTs, done by Bergqvist (16), a member of the Swedish National Board of Occupational Safety and Health, is in full agreement with the US Academy's conclusions.

*Ultraviolet radiation...*

The ultraviolet radiation emitted by the Sun is to a large extent absorbed in the atmosphere. The remaining amount of ultraviolet radiation on the Earth has important beneficial effects on life; it is, for example, indispensable for growth, producing vitamin D in the body. Furthermore, it is responsible for the much-liked and very fashionable sun tan. Excessive doses, however, lead to burns and are potentially carcinogenic.

*... at VDTs is many times lower than standards limits*

*The ultraviolet radiation levels measured in front of VDTs (16, 153) are two or three orders of magnitude below the occupational standards for ultraviolet radiation.* They have been found to be far lower than those of fluorescent lights and thousands of times lower than outdoor sunlight. Thus all experts in that field agree that VDTs do not generate ultraviolet radiation in amounts which could have effects on operators.

*Infrared radiation at VDTs...*

Infrared radiation lies next to visible light in the electromagnetic spectrum. When absorbed by materials or by the human body, infrared radiation is transformed into heat, which usually has a beneficial effect. Infrared radiation is primarily generated by facilities or devices designed to provide heat, such as floor and ceiling heating, stoves, electric heaters or electric blankets.

All measurements of infrared radiation in the environment of a working VDT disclosed emission levels of less than 1% of outdoor or indoor levels.

*... and microwaves are insignificantly low*

Microwaves lie between infrared and radiofrequency waves. They occur as very low cosmic natural background or are generated by radar, satellite communications, FM broadcasting or microwave ovens. In excessively high doses they can have injurious effects; but such levels are found only in foundries and in heavy industry. All measurements of this range of the electromagnetic spectrum at working VDTs showed values which were one thousandth to one billionth the level of the strictest safety regulations.

*Radio-frequency radiation*

The range of radio-frequency radiation is divided into high, low and very low radio frequencies. Excessive doses of the high frequency range can generate thermal effects similar to those of microwaves. Little is known about the effects of low radio frequencies. Radio-frequency radiation is associated with all radio communications and broadcasting, and is found near most electrical appliances.

The horizontal deflection system of the electron beam is a major source of radio frequency radiation in cathode ray tubes. A high voltage circuit accelerates the electrons from the anode, and in order to achieve beam deflection and high voltage generation functions, the horizontal output must be pulsed on and off approximately 15 000—20 000 times per second. This pulsating high voltage creates radiofrequency fields, which are present in relatively minute quantities around VDTs.

*...is similar to levels in urban areas*

The measurements of radiofrequency radiation around VDTs revealed levels comparable to ambient levels generated by radio transmitters in urban areas. They are 2000 times lower than those permitted by the strictest safety regulations.

## 10.2. Electrostatic fields

Electrostatic potentials may occur in front of VDT screens. Bergqvist (16) states that these potentials depend on the devices accelerating the electrons in the direction of the screen. This accelerating potential is applied to the metal-coated inside of the screen and may amount to 20 kV or more. The screen itself should ideally serve as a shield for the field generated by this potential, but such protection may be imperfect, probably due to the limited conductance of the screen itself.

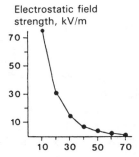

**Figure 99** Electrostatic field strength between the screen and the field mill sensor at measuring distances between 10 and 70 cm in front of the screen.
*According to Knave et al. (105).*

*Electrostatic fields in front of VDTs*

Bergqvist (16) summarized several reports on measurements of electrostatic fields from VDTs, which revealed a wide range between 0 and more than 100 kV/m. The differences were partly due to the different makes, measuring procedures and measuring distances to the screen. *A majority of measurements showed electrostatic fields between 0 and 20 kV/m.*

Bergqvist (16) calculated the screen potentials of the different investigations and categorized the VDTs into three distinct groups — a low group with screen potentials between 0 and about 1 kV, a medium group with values of approximately 5 kV, and a high group with screen potentials around 10 kV.

A Swedish research group (Knave *et al.*, 105) studied the electrostatic phenomena in front of VDT screens in a very systematic way. They determined the electrostatic field at various distances from the screen (from 10 to 70 cm), for all worksites of a large number of VDT operators. The casing of the field mill was earthed and the person taking the measurement was also earthed through the casing. The employee was not sitting at the working place during the measurements.

The measured field strength diminished with increasing distance and was 2–4 kV/m when measured at normal working distances for the group as a whole. These results are shown in Figure 99 and reveal considerable differences between individual makes and screen models. But by and large the measured potentials were in agreement with the results of other authors.

*Electrostatic potential of the operator*

Electrostatic potentials of humans have often been assessed; they depend to a large extent on the humidity of the air as well as on the electric conductance of clothes, shoes and carpets.

The potentials of VDT operators have also been measured; they were negative in many cases, with a mean of $-0.9$ kV (16); with other operators having a potential close to zero or a positive potential with a mean of approximately 1.1 kV. Knave et al. (105) have determined the electrostatic charging of operators. The employee had to hold the un-earthed field mill and point it at an earthed metal disc at a distance of 1 cm. The results revealed a negative field due to an electrostatic charge of about 0.5 kV/m for the VDT operators and of 4 kV/m for the control group.

*The electrostatic field between screen and operator*

Bergqvist (16) made a rough calculation of the electrostatic field between screen and operator. In one study he obtained a field between an 'average' screen and an 'average' operator (at a distance of 60 cm) of $+4.9$ kV/m, and in another study he measured $+10$ kV/m. In another study, Knave et al. (105) estimated fields between 0 and approximately 15 kV/m in work situations.

It is well known that the tiny dust particles in the air may have an electric charge of positive and negative ions. These charged particles are affected by electrostatic fields. The above-mentioned authors measured the positive as well as the negative ions close to the faces of VDT operators, and found that the operators showed higher amounts of ions than control subjects. The calculated particle migration speed was between 0 and 7.5 mm/s, which was considerably lower than the recorded air velocities at VDT workstations.

All these electrostatic phenomena, taken by themselves, certainly have no direct adverse effects on VDT operators, but they may be of relevance for the appearance of skin rashes, which will be discussed below.

## 10.3. Skin rashes

Several authors have reported cases of facial skin rashes, mainly in Scandinavian countries. The first report was published in 1980 by the Norwegian Medical Inspector, Tjonn (198). These skin rashes appeared most frequently around the cheekbones. The affected skin showed redness and some minor desquamation; some cases also exhibited small papulae (small elevations of the skin or skin nodules). The subjects complained of itching and red rash development after between half an hour and several hours' work at the VDT. The rash began to disappear some hours after leaving the place of work.

*Reversible skin rashes may occur over cheek bones*

Tjonn (198) reported 16 cases which he considered to be facial dermatitis related to working conditions. All cases occurred during the winter months. The floors of the offices in which these cases were located were covered with carpets of synthetic fibres, except for one which had a vinyl covering. Some provocation tests (6 hours exposure of one cheek to the screen) were deemed negative, while others disclosed positive results with redness, desquamation and papulae. A window glass between the operator and the screen did not prevent facial rashes. Putting up a glass basin filled with an electrolyte and connected to the surface of the unit and then earthed, seemed to stop the development of rashes. In one office the floor carpeting was replaced by an antistatic one: the skin trouble got worse! It was found that the new carpet had an electrical conductivity close to zero. After replacing that carpet and the isolating glue, the proper earthing was secured and skin irritations and complaints disappeared. Later on Tjonn observed another 20 cases of skin rash in Norway (199) and all these observations led to *his conclusion that some VDT operators may acquire a contact dermatitis related to airborne dust particles and to the electrostatic fields between screen and operators.*

*Dust particles and electrostatic fields are suspected causes*

Rycroft and Calnan (175) reported four cases of skin rash which occurred in a large air-conditioned open-plan office in London. They tried several measures: the change from artificial lighting to daylight had no effect; the use of a skin cream and the positioning of a window glass in front of the screen were both successful in one case. The authors observed a good deal of static electricity derived from nylon carpets. No link was found between the VDTs and facial rashes.

*Self-reported skin disorders*

In the NIOSH study, Smith et al. (154) put items related to "skin rash, itching skin and allergic skin reactions" into the questionnaire. They found a high incidence of self-reported skin troubles at three different sites (44%, 32%, 62%); the differences in occurrence between VDT and non-VDT workers were considerable, but only in one site was this difference significant. It is obvious that *such self-reported skin troubles cannot be compared with the medically diagnosed skin rashes reported in Norway and England.*

Linden and Rolfsen (133) and Nilsen (159) also observed skin disorders among VDT users which they considered occupational dermatitis, i.e., skin inflammations due to contact with substances generating allergic reactions. They suggested that electrostatic fields, low air humidity and carpeting without anti-static qualities might favour the occurrence of skin rashes.

*Knave study: no correlation with electrostatic phenomena*

Knave et al. (104, 105) studied both health complaints and exposure factors, including electrostatic fields at the VDT workplaces. The self-reported skin disorders (skin rashes and irritations on face, neck, hands and arms) indicated a high incidence of about 35% among 395 VDT operators compared to about 25% among 141 control subjects. The analyses of relations between exposure factors and subjective disorders and symptoms as registered in the standardized questionnaires indicated virtually no correlations. In particular, the authors emphasize that they did not find any correlations between electrostatic fields from the screen or from electrostatic charges of the operator and reported skin disorders. Furthermore, the study did not reveal any correlation between the estimated screen—operator field and alleged skin disorders. The former objection must be repeated here: the self-reported skin disorders of the Knave study are certainly not identical with the skin rashes diagnosed by medical doctors, who consider them to be allergic skin dermatitis. Thus it is not surprising that the conclusion of Knave *et al.* (105) does not echo the opinions of Tjonn (199) and others (133, 159), who suggest that the electrostatic field is involved in the aetiology of skin rashes. On the other hand, it must be admitted that the number of medically described cases of facial rash is small and that up to the present these have been observed mainly in the Scandinavian countries. *Nevertheless, the belief that electrostatic phenomena may cause facial skin rashes of VDT operators seems to be a reasonably well-justified hypothesis.*

## 10.4. Alleged cataracts

*The normal prevalence of lens opacities...*

A cataract is a disease of the eye lens, characterized by opacities in its cell structure. Such lens opacity may suffice to degrade the optical image and reduce visual acuity. Opacities of a small degree are not perceived by patients, and such inconsequential opacities are extremely common: it is estimated that about 25% of the general population have lens opacities which do not affect vision. Opacities occur more frequently with increasing age. It is said that about 40% of the people aged between 50 and 64 and as many as 60% of elderly people have opacities that do not affect vision.

*...and cataracts*

Cataracts impairing vision are rather rare; the prevalence of lens opacities with visual deficits is about 7% of the whole population, but only about one in every 2000 people (older than 20 years) find it necessary to eventually undergo surgery.

A certain 'normal' prevalence of incipient cataracts must therefore also be expected among VDT operators: about 20–30% among those who are between 20 and 50 years of age, and 40–50% among those who are over 50 years old. Furthermore, 0.5% per thousand of the VDT operators with incipient cataracts may experience visual deterioration, independent of VDT exposure, eventually necessitating surgical intervention.

There is no clearly defined criterion to ascertain an incipient cataract. The diagnosis is based entirely on the subjective evaluation of the ophthalmologist. This situation makes an anecdotal report on a few cases of incipient cataract rather problematic.

*The causes of cataracts*

The causes of cataracts are essentially unknown, but hereditary and age-dependent degenerative factors must be considered among the most important causes. Some cataracts are related to trauma and metabolic disorders as well as to exposure to external agents, such as high levels of ionizing, ultraviolet, infrared or microwave radiation. However, these familiar causes of cataracts account for only a very small proportion of all cataract cases; the great majority of visually disabling cataracts are associated with ageing.

*The report of Zaret*

At the conference on "Health Hazards of VDTs", held in Loughborough, England in 1980, Zaret (212) reported 10 cases of cataracts which were mostly described as "capsular opacification at the posterior surface of the lens". Eight of the cases worked at VDTs, two were air traffic controllers, and two had been exposed to microwave or X-ray radiation before working at a VDT. For one patient surgery was required, whereas all the others were considered to have incipient cataracts. The ten patients were aged between 27 and 54 years. Zaret (212) believed that working at the VDT was the cause of the diagnosed cataracts.

However, as pointed out earlier, a 7% prevalence of lens opacities is to be considered normal among the population at large. Since many people have lens opacities and many people use VDTs, it is hardly surprising that some VDT users also exhibit lens opacities.

In fact, the recent NIOSH study carried out by Smith *et al.* (185) in the newspaper offices of the *Baltimore Sun* revealed an incidence of small opacities (without diminished visual acuity) of 27.1% among VDT users and 32.2% among non-VDT users. Some 1.4–1.5% of both groups had more serious opacities reducing visual acuity. These results were on the whole confirmed by a survey of the Canadian Labour Congress, quoted by Bergqvist (16).

*No cataractogenic effects of VDTs*

*All authors who have dealt extensively with the problem of alleged cataracts agree that there are neither epidemiologic nor experimental data to support Zaret's suggestion that exposure to VDTs could produce cataracts. Furthermore, it must be pointed out that all cataractogenic electromagnetic radiations (ionizing, ultraviolet, infrared and microwave radiations) are lower at VDT workstations than those derived from normal ambient sources, and well below the occupational and environmental safety standards. There is therefore not the* slightest reason to suspect that VDTs have a cataractogenic effect.

## 10.5. Alleged reproductive hazards due to VDT work

*What are reproductive hazards?*

Reproductive hazards are defined as the risks of genetic effects, birth defects, prematurity and perinatal mortality and spontaneous abortions. It is generally assumed that about 10–20% of all pregnancies end in spontaneous abortion. It is well known that smoking and alcohol consumption increase the rate of abortion. Birth defects have a much lower prevalence of a few percent.

*Anecdotal reports from USA and Canada*

Several anecdotal reports on adverse pregnancy outcomes of VDT operators have been published in the USA and in Canada. The best known observations come from 8 different sources (*Toronto Star* newspaper, two attorneys' offices in Canada, an airport in Quebec, three companies in the USA and one hospital in Vancouver). These prevalence groups of malformations, abortions and other adverse pregnancy outcomes are referred to as 'clusters' because of their statistical appearance in clusters. The observed number of pregnancies was limited (between 5 and 27 per cluster) and the number of reported adverse outcomes varied between 3 and 13 per group, i.e., between 36 and 88%!

Surveys of ionizing and non-ionizing radiation were carried out at the *Toronto Star* and the Surrey Memorial Hospital Vancouver; they revealed very low or undetectable emissions from VDTs.

*Adverse pregnancy outcomes statistically not significant*

Purdham (169) carefully analysed all the anecdotal reports and found that the observed adverse pregnancy outcomes could be explained on the basis of chance only. The relatively small number of cases involved did not allow the assessment of significant prevalences. But Purdham (169) made the point that while chance is a possible explanation it is not necessarily the correct explanation.

Investigations of clusters were therefore not satisfactory, since they lack statistical power. Because of the small percentage of expected cases an extremely large difference between cases and controls would be required to obtain statistical significance. It was calculated that in order to get an 80% probability of detecting a relative risk of 1.4 for spontaneous abortion, 670 pregnancies in the exposed as well as in the control group would be required. This means that 12 000 working female employees, 50% exposed to VDTs and 50% not exposed, would have to be observed for two years.

*The retrospective study by Lee and McNamee*

By order of the British Civil Service Medical Advisory Service, Lee and McNamee (129) carried out a retrospective study of pregnancy outcome among data processing operators. The survey was conducted with a questionnaire and concerned pregnancies ending in the period 1974–82. The exposed group were women who had worked with VDTs for at least 10 hours per week during the three months prior to conception and also during the three months after conception. 803 questionnaires were delivered and the response rate was 72%. From the 576 responding operators, 169 had one or more pregnancies; the exposed group had 55 and the controls 114 pregnancies. The distribution of age at conception, alcohol consumption and smoking habits were very similar for both groups.

*The miscarriage rate of exposed operators corresponded to normal rates*

Of the 55 exposed pregnancies, 8 (14.5%) resulted in spontaneous abortion, whereas only 6 (5.3%) of the controls resulted in miscarriages. Three of the exposed and one of the control pregnancies ended in still-birth. Foetal abnormalities were reported for 10 of the exposed and for 11 of the control pregnancies. As mentioned above, the rate of miscarriages in the normal population lies between 10 and 20%. *The miscarriage rate of the exposed group of 14.5% did not appear unduly high.* It was the rate of the control group, 5.3%, that seemed to be unusually low. The authors discussed several possible explanations for the discrepancy between these two rates. Under-reporting of miscarriages by those not working with VDTs seemed likely. The response rate of 72% was low and could have introduced some selective mechanisms. The authors did not carry out a statistical analysis because the two groups were not selected randomly and because the samples were so small that the difference would not have exceeded the 'chance' of sampling variations.

*The Montreal study by McDonald*

An interesting study on a very large sample of 55 500 women who attended one of 11 Montreal hospitals for a delivery or spontaneous abortion was carried out by McDonald et al. (46) from 1982 to 1984. In a preliminary study in 1982/83 the authors recorded a rate of spontaneous abortion of 18%, while birth defects showed an incidence of a few percent.

The 55 500 women were interviewed about previous pregnancies and also about details of employment during current and past pregnancies. The authors point out that the abortion rate was higher for 'past pregnancies' than for 'current pregnancies'. In 'current pregnancies' only the women who attended a hospital were investigated; early abortions were excluded. Hence the rate of spontaneous abortion was 6.7% in current, but 22.1% in past pregnancies.

*Spontaneous abortions in current pregnancies ...*

The reported spontaneous abortions in *current* pregnancies among the various occupational groups were in the range 6.0–7.4%, only some service and cleaning workers revealed higher rates of 9.1 and 13.0%. For the total number of current pregnancies the mean rate of spontaneous abortions was 6.8%.

A study sample was selected on the basis of a high rate of VDT use (at least 15 hours per week). In this study group 3799 pregnancies occurred with an abortion rate of 6.8%. *This means that the VDT users exhibited a rather low rate of abortion; their rates were very similar to those obtained from the occupational groups from which the non-VDT samples were taken.*

*... do not increase among VDT users*

In another analysis of current pregnancies the VDT users were divided into two groups, one with less than 15 hours use per week and another one with more than 15 hours per week. The results revealed that with longer periods of VDT work the rate of spontaneous abortion was higher. This increase was significant, with $p = 0.01$. The authors suggested that some bias in the responses to the inquiries had led to an exaggeration of the extent of VDT use among those who had experienced spontaneous abortion, and this might explain the apparent trend. Bearing in mind the low overall rate of spontaneous abortion in the five occupational groups studied, McDonald (46) suggested that the balance of evidence was against there being any true association between spontaneous abortions and VDT usage. Bergqvist (16) carefully analysed the results of McDonald's Montreal study and concluded as follows: "It appears that the increase in spontaneous abortion among VDT workers noted in the report of McDonald *et al.* is fully explainable by this error (of a selective bias)".

*A Finnish study revealed...*

In 1985 a case–referent study on birth defects and exposure to VDTs during pregnancy was carried out in Finland by Kurppa *et al.* (119). The interview forms of mothers of 1475 children reporting malformations to the Finnish Register of Congenital Malformation were analysed together with the forms of the same number of paired referents without malformations. Each case mother was time- and area-matched with a referent mother (delivery immediately preceding the case delivery in the same maternity health care district). The retrieval rate of the questionnaires was about 97%. The data covered the period from June 1976 until December 1982. They include 365 defects of the central nervous system (brain), 581 orofacial clefts, 360 defects of the skeleton and 169 cardiovascular malformations. The scrutiny revealed that 111 mothers had worked with VDTs during the first trimester of their pregnancy. Of these, 51 were selected as case mothers (with malformations) and 60 were referents. *The results did*

*...no birth defects due to VDT work*

not indicate a teratogenic risk (risk of birth defects) of VDT operators. These results were well in line with two recent Swedish investigations, neither of which suggested a teratogenic hazard for VDT operators. Bergqvist (16) reports on another study conducted in Sweden jointly by the National Board of Occupational Safety and Health and the National Board of Health and Welfare to investigate the relations between work at VDTs and adverse pregnancy outcome. The authors made use of existing registers to define a study group of 500 mothers with adverse pregnancy outcome, each with two referents. Afterwards these mothers were asked to complete a questionnaire on their work history including use of VDTs. A 'steering group of experts' made an evaluation of the preliminary results of the study. Bergqvist (16) concluded that no indication was found that VDT work leads to increased risks of spontaneous abortion, congenital defects, or perinatal death.

*General conclusions*

Evidently *all scientists who have studied the alleged reproductive hazards of work with VDTs come to the same conclusion, namely that there is, at present, no indication of a risk of miscarriage or birth defects for VDT operators. Several authors, however, point out that occasional clusters of spontaneous abortions or birth defects are to be expected, given the widespread use of VDTs and the varying background level of adverse reproductive outcomes.*

On the other hand, it is well understandable that pregnant VDT operators may worry about alleged adverse effects; despite the reassuring scientific evidence, a certain anxiety may remain among VDT operators.

# 11. Recommendations for VDT Workstations

Several standards for the design of VDT workstations have been published in various countries. Research in this field as well as new developments are still in progress and new views may be expected. Standards should therefore have a tentative character and be easily replaceable. The present recommendations should also be considered with this in mind.

## 11.1. Lighting

*Illumination levels*

For conversational tasks the illumination level should be adapted to the printing quality of the source documents and lie between 300 and 500 lx.

For data-entry tasks the illumination level may be similar to traditional offices and lie between 500 and 700 lx.

*Spatial balance of brightness*

The main luminance contrast at a VDT workstation with bright characters and a dark background occurs between screen and source document. This contrast should not exceed a ratio of 1:10. All other surfaces in the visual environment should have surface luminances between those of the screen and source document; this applies in particular to windows and other bright surfaces in the office.

VDTs with dark characters and bright screens do not raise the problem of excessive contrast ratios in the visual environment.

*Positioning of VDT workstations*

The most effective preventive measure against bright reflections on the screen surface is the adequate positioning of the screen with respect to lights, windows and other bright surfaces. It is important to place the VDT workstation at right angles to the window.

*Appropriate light*

No light source should occur within the visual field of a VDT operator. Light sources behind the operator can produce bright reflections on the screen. Light fixtures directly above the operator can dim the characters on the screen. Thus it is recommended that the lighting fixtures be installed parallel to and on either side of the operator—screen axis. It is advisable to use fixtures which provide a confined downward distribution of light, with a luminous flux angle of 45° to the vertical.

## 11.2. Photometric qualities of VDTs

*Luminances and contrasts*

Displays with bright characters should neither exhibit a too dark nor too bright screen background. A luminance contrast ratio between background and characters of 1:6 is already sufficient for good readability. Given a character luminance of 40–50 cd/m$^2$, a background luminance of 6–8 cd/m$^2$ would be appropriate.

The luminances of the VDT set and the keyboard should lie between that of the screen background and that of the source documents.

*Degree of oscillation of character luminances*

The display must be free of flicker for all operators. As a general rule it can be recommended that the degree of character oscillation be lowered to a level comparable to that of phase-shifted fluorescent tubes. Refresh rates up to 80 or 100 Hz with a phosphor decay time of approximately 10 ms to the 10% luminance level are recommended.

*Character sharpness*

The characters should show sharp edges; no blurred border zone should be perceived. If the blurred border zone is less than 0.3 mm, characters appear to have sharp borders.

Poor sharpness is often due to an insufficient focusing device, adjusted too high character luminance or unsuitable antireflective devices.

*Character stability*

The electronic control of the electron beam must ensure good character stability. Neither drift nor jitter should be perceived by the operators. A luminous dot recorded in the middle of a character bar should not exceed a time-dependent variance of luminance of 20%.

*Reflections on screen surfaces*

All antireflective technologies available on the market today have serious drawbacks: some are associated with a decrease of sharpness and an excessively dark screen background, others are easily soiled. If efficiency is weighed against drawbacks, the quarter-wave coatings and the etching—roughening procedures are to be preferred. Reflected glare on the screen surface should be reduced by 5 to 10 times.

*Size of characters and face*

The range of appropriate character sizes on VDTs is 16 to 25 minutes of visual angle. This means that 3 mm is a suitable height for characters at a viewing distance of 50 cm, and 4.3 mm at 70 cm. The following sizes are recommended:

| | |
|---|---|
| Height of capital letters | 3—4.3 mm |
| Width of characters | 75% of height |
| Distance between characters | 25% of height |
| Space between lines | 100—150% of height |

The spaces between dots should not be visible. Thus a dot matrix of $7 \times 9$ offers better legibility than one of $5 \times 5$.

All symbols should have shapes that are easily distinguished from each other and acceptable as reasonable representations of the symbols concerned.

*Dark versus bright characters*

VDTs with dark characters and bright screen background offer the following advantages: reading conditions similar to printed texts, low contrast ratios to the visual environment and less disturbing reflection on the screen. The drawbacks, at present, are: increased risk of flicker and of small stroke width. Thus a refresh rate of 90 Hz, a phosphor with a decay time of approximately 10 ms to the 10% luminance level and a stroke width of about 0.38 mm are recommended.

## 11.3. Ergonomic design of office furniture and keyboards

*Adjustable workstations*

To reduce constrained postures and physical discomfort the furniture should, in principle, be conceived as flexible as possible. A proper VDT workstation should be adjustable in the following ranges:

| | |
|---|---|
| Keyboard height (middle row to floor) | 70—85 cm |
| Screen centre above floor | 90—115 cm |
| Screen backward inclination to horizontal plane | 88—105° |
| Keyboard (middle row) to table edge | 10—26 cm |
| Screen distance to table edge | 50—75 cm |

A VDT workstation without an adjustable keyboard height and without an adjustable height and distance of the screen is not suitable for continuous work. The controls for adjusting the dimensions of a workstation should be easy to handle, particularly at workstations for rotating shiftwork.

It is nearly impossible for an operator to adjust the workstation dimensions by himself. Another person should be in charge of handling the controls while the operator is working at the VDT.

*Space of legs*

Insufficient space for the legs cause unnatural postures. The space at the level of the knees should be at least 60 cm from the table front and at least 80 cm at the level of the feet.

*The appropriate chair*

A reclined sitting posture is justified since it allows the back muscles to relax and decreases the pressure on the intervertebral discs. The traditional office chairs with relatively small backrests are not suitable for a VDT workstation.

The chair should have a backrest 50 cm long (above the seat surface) and an adjustable inclination. The backrest should have a lumbar support (10–20 cm above seat level) and a slightly concave form at the thoracic level. It should be possible to fix the backrest at any position desired.

*Ergonomic design of keyboards*

At VDTs operators must frequently wait for the response of the computer. Thus flat keyboards, which allow operators to rest forearms and wrists on the desk, and which can be easily shifted on the desk, are to be recommended. The home row of the keyboard should not be higher than 30 mm above the desk.

Recently a new design was proposed, characterized by a splitting of the keyboard into two halves, permitting a more natural position of the hands.

## 11.4. Job design for VDT operators

*The risks of stress...*

Occupational stress has been observed among VDT operators exposed to unexpected computer breakdowns. It is proposed that the computer response time should not exceed 5 s.

Furthermore, some VDT operators engaged in very fragmented, repetitive and monotonous work, a potential of data-entry or data acquisition tasks, may show symptoms of stress.

Therefore VDT work should be designed in such a way that the mental and physical work load is matched to the capabilities of the operatives without demanding either too little or too much of them.

*...should be reduced by an appropriate job design*

New forms of work organization, involving an increase in work variety, and broadening the scope of the work with more responsibility and participation should be taken into account when data-entry and data acquisition jobs are planned. Some data-entry tasks could be broadened by organizing 'mixed activities', where pure data-entry alternates with more demanding tasks.

*Rest pauses*

Rest pauses are a physiological necessity if performance, efficiency and well-being are to be maintained. For most office jobs, including VDT work, it is advisable to divide the daily work into four periods, separated by one rest pause of 10–15 minutes in the morning and one in the afternoon shift and by a lunch break of about 45 minutes at midday.

# References

1 S. R. Abramson, H. L. Mason and H. L. Snyder: The effects of display errors and font styles upon operator performance with a plasma panel. *Proceedings of the Human Factors Society's 27th Annual Meeting*, pp. 28–32 (1983).
2 B. Akerblom: *Standing and Sitting Posture*. Nordiska Bokhandeln, Stockholm (1948).
3 B. Aldman and T. Lewin: *Anthropometric Measurements of Sitting Adults*. Report, Chalmers University of Technology and the University of Gothenburg, Gothenburg, Sweden (1977).
4 B. J. G. Andersson and R. Ortengren: Lumbar disc pressure and myoelectric back muscle activity during sitting. 1. Studies on an office chair. *Scandinavian Journal of Rehabilitation Medicine*, 3, 104–114, 122–127 and 128–135 (1974).
5 R. Arndt: Telephone operator reactions to VDTs. Paper presented at the American Industrial Hygiene Association Conference, Portland, Oregon, USA (1981).
6 D. Bauer and C. R. Cavonius: Improving the legibility of visual display units through contrast reversal. In E. Grandjean and E. Vigliani, eds., *Ergonomic Aspects of Visual Display Terminals*, pp. 137–142. Taylor & Francis, London (1980).
7 D. Bauer, M. Bonacker and C. R. Cavonius: Influence of VDU screen brightness on the visibility of reflected images. *Displays*, 242–244 (1981).
8 D. Bauer, M. Bonacker and C. R. Cavonius: Frame repetition rate for flicker-free viewing of bright VDU screens, *Displays*, January, 31–33 (1983).
9 D. Bauer and C. R. Cavonius: *Untersuchungen zur optimalen Darstellungsart bei Bildschirmgeräten im Büro- und Verwaltungsbereich*. Fachgemeinschaft Büro- und Informationstechnik im VDMA, 6000 Frankfurt 71, Lyonerstrasse 18 (1983).
10 D. Bauer: Causes of flicker at VDUs with bright background and ways of eliminating interference. In E. Grandjean, ed., *Ergonomics and Health in Modern Offices*. Taylor & Francis, London (1984).
11 D. Bauer: Improving VDT workplaces in offices by use of a physiologically optimized screen with black symbols on a light background: basic considerations. Proceedings of Ergodesign 84, *Behaviour and Information Technology*, 3, 363–370 (1984).
12 W. S. Beamon and H. L. Snyder: *An Experimental Evaluation of the Spot Wobble Method of Suppressing Raster Structure Visibility*. Technical Report AMRL-TR-75-63, Wright-Patterson Air Force Base, Ohio (1975).
13 Bell Telephone Laboratories, Inc.: *Video Display Terminals – Preliminary Guidelines for Selection, Installation and Use*.
14 T. Bendix and M. Hagberg: Trunk posture and load on the trapezius muscle whilst sitting at sloping desks. *Ergonomics*, 27, 873–882 (1984).

15 C. Benz, R. Grob and P. Haubner: *Designing VDU Workplaces.* Verlag TÜV Rheinland Köln (1983). Deutsche Ausgabe: Gestaltung von Bildschirm-Arbeitsplätzen.
16 U. O. Bergqvist: Video display terminals and health. A technical and medical appraisal of the state of the art. *Scandinavian Journal of Work Environment and Health*, 10, Suppl. 2 (1984).
17 V. Bhatnager, C. G. Drury and S. G. Schiro: Posture, postural discomfort and performance. *Human Factors*, 27, 189–199 (1985).
18 H. Bouma: Visual reading processes and the quality of text displays. In E. Grandjean and E. Vigliani, eds., *Ergonomic Aspects of Visual Display Terminals*. Taylor & Francis, London (1980).
19 R. M. Boynton, E. J. Rinalducci and C. E. Sternheim: Visibility losses produced by transient adaptational changes in the range from 0.4 to 4000 Foot-Lamberts. *Illuminating Engineering*, 64, 217–227 (1969).
20 U. Bräuninger, E. Grandjean, T. Fellmann and R. Gierer: Lighting characteristics of VDTs. *Proceedings of the Zürich Seminar on Digital Communication*, 9–11 March (1982).
21 U. Bräuninger: Lichttechnische Eigenschaften der Bildschirmgeräte aus ergonomischer Sicht. Dissertation, ETH, Zürich, Nr 7310 (1983).
22 U. Bräuninger, E. Grandjean, G. van der Heiden, K. Nishiyama and R. Gierer: Lighting characteristics of VDTs from an ergonomic point of view. In E. Grandjean, ed., *Ergonomics and Health in Modern Offices*. Taylor & Francis, London (1984).
23 D. E. Broadbent: Effects of noise on behaviour. In C. M. Harris, ed., *Handbook of Noise Control.* McGraw-Hill, New York (1957).
24 C. R. Brown and D. L. Schaum: User adjusted VDT parameters. In E. Grandjean and E. Vigliani, eds., *Ergonomic Aspects of Visual Display Terminals*. Taylor & Francis, London (1980).
25 US Bureau of Radiological Health: *An Evaluation of Radiation Emission from Video Display Terminals*. Dept. of Health and Human Services Publication No. FDA 81-8153, Washington, DC (1981).
26 J. Buesen: Product development of an ergonomic keyboard. Proceedings of Ergodesign 84, *Behaviour and Information Technology*, 3, 387–390 (1984).
27 A. Cakir, H. J. Reuter, L. von Schmude and A. Armbruster: *Anpassung von Bildschirmarbeitsplätzen an die physische und psychische Funktionsweise des Menschen*. Bundesministerium für Arbeit und Sozialordnung, Referat Presse, Postfach, Bonn (1978).
28 A. Cakir, D. J. Hart and T. F. M. Stewart: *The VDT Manual.* IFRA, Washingtonplatz 1, Darmstadt (1979)
29 A. Cakir, R. P. Franke and M. Piruzram: *Arbeitsplätze für Phonotypistinnen*. Bundesanstalt für Arbeitsschutz, Report No. 363, Dortmund (1983).
30 S. Cantoni, D. Colombini, E. Occhipinti, A. Grieco, C. Frigo and A. Pedotti: Postural analysis and evaluation at the old and new workplace in a telephone company. In E. Grandjean, ed., *Ergonomics and Health in Modern Offices*. Taylor & Francis, London (1984).
31 R. D. Caplan, S. Cobb, I. R. P. French, R. van Harrison and S. R. Pinneua, Jr.: *Job Demands and Worker Health*. NIOSH Report No. 75–160. National Institute for Occupational Safety and Health, Cincinnati (1975).

32 C. Cavonius: Changes in contrast and sensitivity from VDT usage. Paper presented at the IX International Ergophthalmological Symposium, San Francisco (1982).
33 D. B. Chaffin: Localized muscle fatigue — definition and measurement. *Journal of Occupational Medicine*, 15, 346–354 (1973).
34 D. B. Chaffin and G. Andersson: *Occupational Biomechanics*. John Wiley & Sons, New York (1984).
35 J. B. Coe, K. Cuttle, W. C. McClellan and N. J. Warden: *Visual Display Units. A Review of the Potential Health Problems Associated with Their Use*, Wellington, NZ: Regional Occupational Health Unit, New Zealand Dept. of Health.
36 G. F. Cohen, ed.: *Human Aspects in Office Automation*, Elsevier Series in Office Automation 1, Elsevier, New York (1984).
37 J. B. Collins: The role of a sub-harmonic in the wave-form of light from a fluorescent lamp in causing complaints of flicker. *Ophthalmologica*, 131, 377–387 (1956).
38 E. A. Cox: Radiation emissions from visual display units. In *Health Hazards of VDUs?* HUSAT Research Group, Loughborough University, UK (1980).
39 T. Cox: The nature and measurement of stress. *Ergonomics*, 28, 1155–1163 (1985).
40 W. H. Cushman: Data entry performance and operator preferences for various keyboard heights. In E. Grandjean, ed., *Ergonomics and Health in Modern Offices*. Taylor & Francis, London (1984).
41 M. Dainoff, A. Happ and P. Crane: Visual fatigue and occupational stress in VDT operators. *Human Factors* 23, 421–438 (1981).
42 M. J. Dainoff: Occupational stress factors in VDTs operation: a review of empirical research. *Behaviour and Information Technology*, 1, 141–176 (1982).
43 N. Diffrient, A. R. Tilley and J. C. Bardagjy with Henry Dreyfuss Ass.: *Human scale 1/2/3*. MIT Press, Cambridge, MA (1974).
44 DIN Norm 5035: *Innenraumbeleuchtung mit künstlichem Licht.* Beuth, Berlin and Köln (1972).
45 DIN Norm 4549: *Schreibtische, Büromaschinentische und Bildschirmarbeitstische.* Beuth, Berlin (1981).
46 D. McDonald, N. M. Cherry, C. Delorme and J. C. McDonald: Work and pregnancy in Montreal — preliminary findings on work with VDTs. In *Allegations of Reproductive Hazards from VDUs.* Humane Technology and HUSAT Research Centre, Loughborough University UK (1984).
47 C. G. Drury and M. Francher: Evaluation of a forward-sloping chair, *Applied Ergonomics*, 16, 41–47 (1985).
48 A. Dubois-Poulsen: Notions de physiologie ergonomique de l'appareil visuel. In J. Scherrer, ed., *Physiologie du Travail*, Vol. 2, pp. 114–183, Masson, Paris (1967).
49 K. Dugas Garcia and W. W. Wierwille: Effect of glare on performance of a VDT reading-comprehension task. *Human Factors*, 27, 163–173 (1985).
50 J. Duncan and D. Ferguson: Keyboard operating posture and symptoms in operating. *Ergonomics*, 17 651–662 (1974).
51 R. Elias and F. Cail: Exigences visuelles et fatigue dans deux types de tâches informatisées. *Le Travail Humain*, 46, 81–92 (1983).
52 D. S. Ellis: Speed of manipulative performance as a function of worksurface height. *Journal of Applied Psychology*, 35, 289–296 (1951).

53 F. L. Engel: Information selection from visual display units. In E. Grandjean and E. Vigliani, eds., *Ergonomic Aspects of Visual Display Terminals*. Taylor & Francis, London (1980).

54 U. T. Eysel and U. Burandt: Fluorescent tube light evokes flicker responses in visual neurons. *Vision Research*, 24, 943–948 (1984).

55 T. Fellmann, U. Bräuninger, R. Gierer and E. Grandjean: An ergonomic evaluation of VDTs. *Behaviour and Information Technology*, 1, 69–80 (1982).

56 G. C. Fortuin: *Visual Power and Visibility*. Philips Research Report, Eindhoven (1957).

57 M. Frankenhäuser, B. Nordheden, A. L. Myrsten and B. Post: Psychophysiological reactions to understimulation and overstimulation. *Acta Psychologica*, 35, 298–308 (1971).

58 M. Frankenhäuser: *Man in Technological Society: Stress, Adaptation and Tolerance Limits*. Report from the Psychological Laboratories, University of Stockholm, Suppl. 26 (1974).

59 G. A. Fry and V. M. King: The pupillary response and discomfort glare. *Journal of the IES*, 307–324 (1975).

60 B. Gardell and G. Johansson, eds.: *Working Life: A Social Science Contribution to Work Reform*. John Wiley and Sons, Chichester (1981).

61 L. Ghiringhelli: Collection of subjective opinions on use of VDTs. In E. Grandjean and E. Vigliani, eds., *Ergonomic Aspects of Visual Display Terminals*. Taylor & Francis, London (1980).

62 F. E. Gomer and K. G. Bish: Evoked potential correlates of display image quality. *Human Factors*, 20, 589–596 (1978).

63 J. D. Gould: Visual factors in the design of computer-controlled CRT displays. *Human Factors*, 10, 359–376 (1968).

64 J. D. Gould and N. Grischkowsky: Doing the same work with hard copy and with cathode-ray tube (CRT) computer terminals. *Human Factors*, 26, 323–337 (1984).

65 E. Grandjean, B. Horisberger, L. Havas and K. Abt: Arbeitsphysiologische Untersuchungen mit verschiedenen Beleuchtungs-systemen an einer Feinarbeit. *Industrielle Organisation*, 28, 231–239 (1959).

66 E. Grandjean and H. U. Burandt: Körpermasse der Belegschaft eines schweizerischen Industriebetriebes. *Industrielle Organisation*, 31, 239–242 (1962).

67 E. Grandjean and H. U. Burandt: Das Sitzverhalten von Buroangestellten. *Industrielle Organisation*, 31, 243–250 (1962).

68 E. Grandjean, A. Böni and H. Kretschmar: The development of a rest chair profile for healthy and notalgic people. In E. Grandjean, ed., *Sitting Posture*. Taylor & Francis, London (1969).

69 E. Grandjean, W. Hünting, G. Wotzka and R. Schärer: An ergonomic investigation of multipurpose chairs. *Human Factors*, 15, 247–255 (1973).

70 E. Grandjean: *Fitting the Task to the Man*. Taylor & Francis, London (1980).

71 E. Grandjean, M. Nakaseko, W. Hünting and T. Läubli: Ergonomische Untersuchungen zur Entwicklung einer neuen Tastatur für Büromaschinen. *Zeitschrift für Arbeitswissenschaft*, 35, 221–226 (1981).

72 E. Grandjean, K. Nishiyama, W. Hünting and M. Piderman: A laboratory study on preferred and imposed settings of a VDT workstation. *Behaviour and Information Technology*, 1, 289–304 (1982).

73 E. Grandjean, W. Hünting and M. Pidermann: VDT workstation design: preferred settings and their effects. *Human Factors*, 25, 161–175 (1983).
74 J. P. de Groot and A. Kamphuis: Eyestrain in VDU users: physical correlates and long-term effects. *Human Factors*, 25, 409–413 (1983).
75 E. Gunnarsson and I. Soederberg: Eye strain resulting from VDT work at the Swedish Telecommunications Administration. *Applied Ergonomics*, 14, 61–69 (1983).
76 S. K. Guth: Light and comfort. *Industrial Medicine and Surgery*, 27, 570–574 (1958).
77 S. Gyr, K. Nishiyama, R. Gierer, T. Läubli and E. Grandjean: The effect of various refresh rates in positive and negative displays. In E. Grandjean, ed., *Ergonomics and Health in Modern Offices*. Taylor & Francis, London (1984).
78 I. K. Habinek, P. M. Jacobson, W. Miller and T. W. Suther: A comparison of VDT antireflection treatments. In *Proceedings of the 26th Annual Meeting of the Human Factors Society*, pp. 285–289. Human Factors Society, Santa Monica, CA (1982).
79 M. Hagberg: *Arbetsrelaterade Besvär i Halsrygg och Skuldra*. Swedish Work Environment Fund Report, Stockholm (1982).
80 P. Haubner and S. Kokoschka: Bewertung verschiedener Anti-reflexmassnahmen für Bildschirmgeräte. *Lichtforschung*, 2, 91–93 (1980).
81 P. Haubner and S. Kokoschka: *Visual Display Units – Characteristics of Performance*. International Commission on Illumination (CIE), 20th Session in Amsterdam, 52 Bd Malesherbes, Paris (1983).
82 L. R. Hedman and V. Briem: Short term changes in eyestrain of VDU users as a function of age. *Human Factors*, 26, 357–370 (1984).
83 G. van der Heiden and H. Krueger: *Evaluation of Ergonomic Features of the Computervision Instaview Graphics Terminal*. Report of the Dept. of Ergonomics, Swiss Federal Institute of Technology, Zürich (1984).
84 G. van der Heiden, U. Bräuninger and E. Grandjean: Ergonomic studies on computer aided design. In E. Grandjean, ed., *Ergonomics and Health in Modern Offices*. Taylor & Francis, London (1984).
85 G. van der Heiden: Ergonomische Anforderungen an Arbeits-plätze für Computer Aided Design (CAD). Thesis, Swiss Federal Institute of Technology, Zürich (1986).
86 M. G. Helander, P. A. Billingsley and J. M. Schurik: An evaluation of human factors research on VDTs in the work-place. *Human Factors Review*, 55–129 (1984).
87 S. G. Hill and K. Kroemer: Preferred declination of the line of sight. Unpublished.
88 P. A. Howarth and H. O. Istance: The association between visual discomfort and the use of visual display units. *Behaviour and Information Technology*, 4, 131–149 (1985).
89 W. C. Howell and C. L. Kraft: The judgement of size, contrasts and sharpness of letter forms. *Journal of Experimental Psychology*, 61, 30–39 (1961).
90 E. Hort: A new concept in chair design, Proceedings of Ergodesign 84, *Behaviour and Information Technology*, 3, 359–363 (1984).
91 W. Hünting and E. Grandjean: Sitzverhalten und subjektives Wohlbefinden auf schwenkbaren und fixierten Formsitzen, *Zeitschrift für Arbeitswissenschaft*, 30, 161–164 (1976).

92 W. Hünting, E. Grandjean and K. Maeda: Constrained postures in accounting machine operators. *Applied Ergonomics*, 11, 145–149 (1980).

93 W. Hünting, T. Läubli and E. Grandjean: Postural and visual loads at VDT workplaces, Part 1: constrained postures. *Ergonomics*, 24, 917–931 (1981).

94 W. Hünting, M. Nakaseko, R. Gierer and E. Grandjean: Ergonomische Gestaltung von alphanumerischen Tastaturen. *Sozial- und Präventivmedizin*, 27, 251–252 (1982).

95 IES: *IES Lighting Handbook*, 5th ed. Illuminating Engineering Society, New York (1972).

96 IBM: *Human Factors of Workstations with Visual Displays*. IBM Human Factors Centre, 5600 Cottle Road, San José, CA.

97 S. H. Isensee and C. A. Bennett: The perception of flicker and glare on computer CRT displays. *Human Factors*, 25, 177–184 (1983).

98 G. Johansson, G. Aronsson and B. O. Lindström: *Social Psychological and Neuroendocrine Stress Reactions in Highly Mechanized Work*. Report from the Psychological Laboratories, University of Stockholm, No. 488 (1976).

99 G. Johansson and G. Aronsson: *Stress Reactions in Computerized Administrative Work*. Reports from the Department of Psychology, University of Stockholm, Suppl. 50, November (1980).

100 J. J. Keegan: Alterations of the lumbar curve related to posture and seating. *Journal of Bone-Joint Surgery*, 35, 567–589 (1953).

101 E. C. Keighley and P. H. Parkin: Subjective response to sound conditioning in a landscaped office. *Journal of Sound and Vibration*, 64, 313–323 (1979).

102 D. H. Kelly: Visual response to time-dependent stimuli. *Journal of the Optical Society of America*, 51, 422–429 (1961) and 52, 89–95 (1962).

103 E. A. Klockenberg: *Rationalisierung der Schreibmaschine und ihrer Bedienung*. Springer, Berlin (1926).

104 B. G. Knave, R. I. Wibom, M. Voss, L. D. Hedström and U. O. Bergqvist: Work with VDTs among office employees. I. Subjective symptoms and discomfort. *Scandinavian Journal of Work Environment and Health*, 11, 457–466 (1985).

105 B. G. Knave, R. I. Wibon, U. O. Bergqvist, L. Carlsson, M. Levin and P. Nylen: Work with VDTs among office employees. II. Physical exposure factors. *Scandinavian Journal of Work Environment and Health*, 11, 467–474 (1985).

106 S. Kokoschka and H. W. Bodmann: Untersuchungen zum Beleuchtungs-niveau und Zeichenkontrast am Bildschirmarbeitsplatz. *Lichttechnik*, 30, 395–399 (1978).

107 Y. Komoike and S. Horiguchi: Fatigue assessment on key punch operators, typists and others. *Ergonomics*, 14, 101–109 (1971).

108 D. Korell: Planning for the future office — today. Proceedings of Ergodesign 84. *Behaviour and Information Technology*, 3, 329–340 (1984).

109 A. Korge and H. Krueger: Influence of edge sharpness on the accommodation of the human eye. *Graefe's Archive of Clinical Experimental Ophtalmology*, 222, 26–28 (1984).

110 J. Krämer: *Biomechanische Veränderungen im lumbalen Bewegungssegment*. Hippokrates, Stuttgart (1973).

111 K. H. E. Kroemer: Heute zutreffende Körpermasse. *Zeitschrift für Arbeitswissenschaft*, 3, 42–45 (1964).

112 K. H. E. Kroemer: Über den Einfluss der räumlichen Lage von Tastenfeldern auf die Leistung an Schreibmaschinen. *Internationale Zeitschrift für angewandte Physiologie*, 20, 453–464 (1965).

113 K. H. E. Kroemer: Human engineering — the keyboard. *Human Factors*, 14, 51–63 (1972).
114 H. Krueger and W. Müller-Limmroth: *Arbeiten mit dem Bildschirm — aber richtig!* Bayerisches Staatsministerium für Arbeit und Sozialordnung, Winzererstr 9, Munich (1979).
115 H. Krueger and R. Mader: Der Einfluss der Farbsättigung auf den chromatischen Fehler der Akkomodation des menschlichen Auges. *Fortschritte der Ophthalmologie*, 79, 171–173 (1982).
116 H. Krueger and J. Hessen: Objektive kontinuierliche Messung der Refraktion des Auges. *Biomedizinische Technik*, 27, 142–147 (1982).
117 H. Krueger: *Zur Ergonomie von Balans-Sitzelementen in Hinblick auf ihre Verwendbarkeit als reguläre Arbeitsstühle.* Report of the Dept. of Ergonomics, Swiss Federal Institute of Technology, Zürich (1984).
118 R. Kruk and P. Mutter: Reading of continuous text on video screens. *Human Factors*, 26, 339–345 (1984).
119 K. Kurppa, P. C. Holmberg, K. Rantala, T. Nurminen and L. Saxen: Birth defects and exposure to VDTs during pregnancy. *Scandinavian Journal of Work Environment and Health*, 11, 353–356 (1985).
120 T. Läubli, W. Hünting and E. Grandjean: Postural and visual loads at VDT workplaces, Part 2: lighting conditions and visual impairments. *Ergonomics*, 24, 933–944 (1981).
121 T. Läubli: *Das arbeitsbedingte cervicobrachiale Überlastungssyndrom.* Thesis, Medical Faculty, University of Zürich (1981).
122 T. Läubli, W. Hünting, E. Grandjean, T. Fellmann, U. Bräuninger and R. Gierer: Belastungsfaktoren an Bildschirmarbeitsplätzen, *Klinische Monatsblätter für Augenheilkunde*, 180, 363–366 (1982).
123 T. Läubli and E. Grandjean: The magic of control groups in VDT field studies. In E. Grandjean, ed., *Ergonomics and Health in Modern Offices*. Taylor & Francis, London (1984).
124 T. Läubli, H. Mion, E. Senn, C. Thomas and H. Zeier: Rheumatische Beschwerden und Büroarbeit. *Sozial- und Präventivmedizin*, 30, 278–279 (1985).
125 T. Läubli, E. Senn, W. Fasser, H. Mion, T. Carlo and H. Zeier: Klinische Befunde und subjektive Klagen über Beschwerden im Bewegungsapparat. *Sozial- und Präventivmedizin*, 31, (in press) (1986).
126 T. Läubli, K. Nishiyama, R. Gierer and E. Grandjean: Effects on refresh rates of a simulated CRT display with bright characters on a dark screen. *International Journal of Industrial Ergonomics*, 1, (in press) (1986).
127 M. Launis: Design of a VDT workstation for customer service. In E. Grandjean, ed., *Ergonomics and Health in Modern Offices*. Taylor & Francis, London (1984).
128 A. Laville: Postural reactions related to activites on VDUs. In E. Grandjean and E. Vigliani, eds., *Ergonomic Aspects of Visual Display Terminals*. Taylor & Francis, London (1980).
129 B. V. Lee and R. McNamee: Reproduction and work with VDUs — a pilot study. In *Allegations of Reproductive Hazards from VDUs*. Humane Technology and HUSAT Research Centre, Loughborough University, UK (1984).
130 G. Lehmann and F. Stier: Mensch und Gerät. *Handbuch der gesamten Arbeitsmedizin*, Band 1, pp. 718–788. Urban & Schwarzenberg, Berlin (1961).
131 L. Levi: *Emotions — Their Parameters and Measurement*. Raven Press, New York (1975).

132 N. A. Life and S. T. Pheasant: An integrated approach of the study of posture in keyboard operation. *Applied Ergonomics*, 15, 83–90 (1984).
133 V. Linden and S. Rolfsen: Video computer terminals and occupational dermatitis. *Scandinavian Journal of Work Environment and Health*, 7, 62–64 (1981).
134 H. Luckiesh and F. K. Moss: *The Science of Seeing*. Van Nostrand, New York (1937).
135 A. Lundervold: Electromyographic investigations of position and manner of working in typewriting. *Acta Physiologica Scandinavia*, 24, Suppl. 84, 1–171 (1951).
136 A. Lundervold: Electromyographic investigations during typewriting. *Ergonomics*, 1, 226–233 (1958).
137 K. Maeda, W. Hünting and E. Grandjean: Localized fatigue in accounting-machine operators, *Journal of Occupational Medicine*, 22, 810–816 (1980).
138 K. Maeda, S. Horiguchi and M. Hosokawa: History of the studies on occupational cervicobrachial disorder in Japan and remaining problems. *Journal of Human Ergology*, 11, 17–29 (1982).
139 A. C. Mandal: What is the correct height of furniture? In E. Grandjean, ed., *Ergonomics and Health in Modern Offices*. Taylor & Francis, London (1984).
140 M. Menozzi and H. Krueger: Computerized method for analyzing display quality. Paper presented at the Conference "Work with Display Units", Stockholm (1986).
141 J. J. Meyer, R. Gramoni, S. Korol and P. Rey, Quelques aspects de la charge visuelle aux postes de travail impliquant un écran de visualisation, *Le Travail Humain*, 42, 275–301 (1979).
142 I. Miller and T. W. Suther: Preferred height and angle setting of CRT and keyboard for a display station input task. *Proceedings of the Human Factors Society's 25th Annual Meeting*. Santa Monica, CA, Human Factors Society (1981).
143 G. Molteni, A. Grieco, D. Colombini, E. Occhipinti, A. Pedotti, S. Boccardi, G. Frigo and O. Menoni: Analisi delle posture. *C.E.E.*, 6, 6 (1983).
144 H. Monod: Contributions à l'Étude du Travail Statique. Thése, Faculté de Medicine, Paris (1956).
145 C. T. Morgan, A. Chapanis, J. S. Cook and M. W. Lund: *Human Engineering Guide to Equipment Design*, McGraw-Hill, New York (1963).
146 R. R. Mourant, R. Lakshmanan and R. Chantadisai: Visual fatigue and CRT display terminals. *Human Factors*, 23, 529–540 (1981).
147 K. F. H. Murrell: *Ergonomics; Man in His Working Environment*, Chapman and Hall, London (1965).
148 P. Mutter, S. Latremouille, W. Treurniet and P. Beam: Extended reading of continuous text on television screens. *Human Factors*, 24, 501–508 (1982).
149 A. Nachemson and G. Elfström: Intravertebral dynamic pressure measurements in lumbar discs, *Scandinavian Journal of Rehabilitation Medicine*, Suppl. 1 (1970).
150 N. Nakaseko, E. Grandjean, W. Hünting and R. Gierer: Studies on ergonomically designed alphanumeric keyboards. *Human Factors*, 27, 175–187 (1985).
151 Choon Nam Ong: VDT workplace design and physical fatigue. In E. Grandjean, ed., *Ergonomics and Health in Modern Offices*. Taylor & Francis, London (1984).

153 National Academy Press: *Video Displays, Work and Vision.* Report of the panel on impact of video viewing on vision of workers. NAP, Washington, DC (1983).
154 National Institute for Occupational Safety and Health: *Potential Health Hazards of VDTs.* Research Report No. 81–129, US Government Printing Office, Washington, DC (1981).
155 J. Nemecek and E. Grandjean: Das Grossraumbüro in arbeitsphysiologischer Sicht, *Industrielle Organisation,* 40, 233–243 (1971).
156 J. Nemecek and N. Turrian: Der Bürolärm und seine Wirkungen. *Kampf dem Larm,* 25, 50–57 (1978).
157 F. L. Van Nes: New characters for teletext with improved legibility. *Institute for Perception Research, Annual Progress Report,* 18, 108–113 (1983).
158 F. L. Van Nes: Limits of visual perceptions in the technology of visual display terminals. In Proceedings of Ergodesign 84, *Behaviour and Information Technology,* 3, 371–377 (1984).
159 A. Nilsen: Facial rash in VDU operators. *Contact Dermatitis,* 8, 25–28 (1982).
160 K. Nishiyama, M. Nakaseko and T. Uehata: Health aspects of VDT operators in the newspaper industry. In E. Grandjean, ed., *Ergonomics and Health in Modern Offices.* Taylor & Francis, London (1984).
161 K. Nishiyama, U. Bräuninger, H. de Boer, R. Gierer and E. Grandjean: Physiological effects of intermittently illuminated textual displays. *Ergonomics,* 29, 1145–1156 (1986).
162 K. G. Nyman, B. G. Knave and M. Voss: Work with VDTs among office employees. IV. Refraction, accommodation, convergence and binocular vision. *Scandinavian Journal of Work and Environmental Health,* 11, 483–487 (1985).
163 E. Occhipinti, D. Colombini, C. Frigo, A. Pedotti and A. Grieco: Sitting posture: analysis of lumbar stresses with upper limbs supported. *Ergonomics,* 28, 1333–1346 (1985).
164 O. Östberg: Health problems for operators working with CRT displays. *International Journal of Occupational Health and Safety,* Nov/Dec, 24–52 (1975).
165 O. Östberg, J. Powell and A. Blomquist: *Laser Optometry in Assessment of Visual Fatigue in VDU Operators.* University of Lulea, Sweden, Technical Report 1T (1980).
166 O. Östberg, G. Ahlström and L. Möller: Ergonomic procurement guidelines for VDUs as a tool for progressive change. *Proceedings of 11th International Symposium on Human Factors in Telecommunications,* 9–13 Sept. (1985).
167 A. M. Paci: Contribution of ergonomics to the design of antireflection devices in the development of VDU work-places. Proceedings of Ergodesign 84, *Behaviour and Information Technology,* 3, 381–386 (1984).
167b S. T. Pheasant: *Bodyspace: Anthropometry, Ergonomics and Design.* Taylor & Francis (1986).
168 *Proceedings of the International Conference on Enhancing the Quality of Working Life.* Arden House, Harriman, New York (1972).
169 J. Purdham: Adverse pregnancy outcome amongst VDT operators — the cluster phenomenon. In *Allegations of Reproductive Hazards from VDUs.* Humane Technology and HUSAT Research Centre, Loughborough University, UK (1984).

170 G. W. Radl: Experimental investigations for optimal presentation—mode and colours of symbols on the CRT-screen. In E. Grandjean and E. Vigliani, eds., *Ergonomic Aspects of Visual Display Terminals*, pp. 127–137. Taylor & Francis, London (1980).

171 E. J. Rinalducci and A. N. Beare: Losses in night-time visibility caused by transient adaptation. *Journal of the Illuminating Engineering Soceity*, 3, 336–345 (1974).

172 T. Rubin and C. J. Marshall: Adjustable VDT workstations: can naive users achieve a human factors solutions? In *International Conference on Man-Machine Systems*, IEE Conf. Publ. No. 121, Manchester (1982).

173 B. A. Rupp: Visual display standards: a review of issues. *Proceedings of the Society for Information Display*, 22, 63–72 (1981).

174 B. A. Rupp, B. W. McVey and S. E. Taylor: Image quality and the accommodation response. In E. Grandjean, ed., *Ergonomics and Health in Modern Offices*, Taylor & Francis, London (1984).

175 R. J. G. Rycroft and C. D. Calnan: Facial rashes among visual display unit operators. In B. G. Pearce, ed., *Health Hazards of VDTs?* John Wiley & Sons, Chichester, UK (1984).

176 G. Salvendy: Research issues in the ergonomics, behavioural, organizational and management aspects of office automation. In G. F. Cohen, ed., *Human Aspects in Office Automation*. Elsevier, New York (1984).

177 S. L. Sauter, M. S. Gottlieb, K. C. Jones, V. N. Dodson and K. M. Rohrer: Job and health implications of VDT use: initial results of the Wisconsin–NIOSH study. *Communications of the ACM*, 26, 284–294 (1983).

178 S. L. Sauter: Predictors of strain in VDT users and traditional office workers. In E. Grandjean, ed., *Ergonomics and Health in Modern Offices*. Taylor & Francis, London (1984).

179 H. Schoberth: *Sitzhaltung – Sitzschaden – Sitzmöbel*. Springer, Berlin (1962).

180 H. Shahnavaz: Lighting conditions and workplace dimensions of VDU operators. *Ergonomics*, 25, 1165–1173 (1982).

181 D. A. Shurtleff: *How to Make Displays Legible*. Human Interface Design, La Mirada, CA (1980).

182 S. J. Shute and S. J. Starr: Effects of adjustable furniture on VDT users. *Human Factors*, 26, 157–170 (1984).

183 A. B. Smith, S. Tanaka, W. Halperin and R. D. Richards: *Report of a Cross-Sectional Survey of VDT Users at the Baltimore Sun*. National Institute for Occupational Safety and Health, Centre for Disease Control, Cincinnati (1982).

184 A. B. Smith, S. Tanaka and W. Halperin: Correlates of ocular and somatic symptoms among VDT users. *Human Factors*, 26, 143–156 (1984).

185 M. J. Smith, L. W. Stammerjohn, G. F. Cohen and N. R. Lalich: Job stress in video display operations. In E. Grandjean and E. Vigliani, eds., *Ergonomic Aspects of Visual Display Terminals*. Taylor & Francis, London, (1980).

186 M. J. Smith, B. G. F. Cohen, L. W. Stammerjohn and A. Happ: An investigation of health complaints and job stress in video display operations. *Human Factors*, 23, 387–399 (1981).

187 H. L. Snyder and M. E. Maddox: *Information Transfer from Computer Generated Dot Matrix Displays*. Technical report HFL-78-3, Virginia Polytechnic Institute and State University, Virginia (1978).

188 H. L. Snyder and M. E. Maddox: On the image quality of dot-matrix displays. *Proceedings of the SID*, 21, 3–7 (1980).

189 H. L. Snyder: *Human Visual Performance and Flat Panel Display Image Quality*. Technical Report HFL-80-1, Virginia Polytechnic Institute and State University, Virginia (1980).
190 H. L. Snyder: Lighting characteristics, legibility and visual comfort at VDTs. In E. Grandjean, ed., *Ergonomics and Health in Modern Offices*, Taylor & Francis, London (1984).
191 L. W. Stammerjohn, M. J. Smith and B. G. F. Cohen: Evaluation of workstation design factors in VDT operations. *Human Factors*, 23, 401–412 (1981).
192 S. J. Starr, C. R. Thompson and S. J. Shute: Effects of video display terminals on telephone operators. *Human Factors*, 24, 699–711 (1982).
193 S. J. Starr: Effects of VDTs in a business office. *Human Factors*, 26, 347–356 (1984).
194 T. F. M. Stewart: Eyestrain and visual display units: a review. *Displays*, April (1979).
195 T. F. M. Stewart: The role of ergonomics in office-systems design. Proceedings of Ergodesign 84, *Behaviour and Information Technology*, 3, 319–328 (1984).
196 T. Terrana, F. Merluzzi and E. Giudici: Electromagnetic radiations emitted by VDUs. In E. Grandjean and E. Vigliani, eds., *Ergonomic Aspects of Visual Display Terminals*. Taylor & Francis, London (1980).
197 H. Timmers: An effect of contrast on legibility of printed text. *IPO Annual Progress Report No. 13*, 64–67 (1978).
198 H. Tjonn: Can VDU operation cause dermatitis? In E. Grandjean and E. Vigliani, eds., *Ergonomic Aspects of Visual Display Terminals*, Taylor & Francis, London (1980).
199 H. Tjonn: Report of facial rashes among VDU operators in Norway. In B. Pearce, ed., *Health Hazards of VDUs?* John Wiley & Sons, Chichester, UK (1984).
200 E. Ulich, P. Groskurth and A. Bruggemann: *Neue Formen der Arbeitsgestaltung*. Europäische Verlagsanstalt, Frankfurt (1973).
201 US Dept of Health, Education and Welfare: Weight, height and selected body dimensions of adults, *United States 1960–62 Vital and Health Statistics*, 11 (8), Washington: Government Printing Office (1966).
202 A. Vartabedian: Legibility of symbols on CRT displays, *Applied Ergonomics*, 2, 130–132 (1971).
203 Verwaltungs- und Berufsgenossenschaft: *Sicherheitsregeln für Bildschirmarbeitsplätze im Bürobereich*. Überseering 8, Hamburg (1981).
204 B. W. McVey, C. K. Clauer and S. E. Taylor: A comparison of anti-glare contrast-enhancement filters for positive and negative image displays under adverse lighting conditions, in E. Grandjean, ed., *Ergonomics and Health in Modern Offices*, pp. 405–409. Taylor & Francis, London (1984).
205 S. Webb and R. Coburn: *Development and Testing of a Hand-configured Keyset*. Technical Memorandum No. TM-357, US Navy Electronics Lab., San Diego (1959).
206 A. Weber, E. Sancin and E. Grandjean: The effects of various keyboard heights on EMG and physical discomfort. In E. Grandjean, ed., *Ergonomics and Health in Modern Offices*. Taylor & Francis, London (1984).
207 M. W. Weiss and R. C. Petersen: Electromagnetic radiation emitted from video computer terminals, *American Industrial Hygiene Association Journal*, 40, 300–309 (1979).
208 P. van Wely: Design and disease. *Applied Ergonomics*, 1, 262–269 (1970).

209 R. Wiebelitz and H. Schmitz: Flimmerndes Licht und die menschliche Augenpupille. *Zeitschrift für Arbeitswissenschaft*, **37**, 163–168 (1983).
210 A. Wisner and R. Rebiffé: L'utilisation des données anthropometriques dans la conceptions du poste de travail. *Travail Humain*, **26**, 193–217 (1963).
211 Y. Yamaguchi, F. Umezawa and Y. Ishinada: Sitting posture: an electromyographic study on healthy and notalgic people. *Journal of the Japanese Orthopaedic Association*, **46**, 51–56 (1972).
212 M. M. Zaret: Cataracts and VDUs. In B. G. Pearce, ed., *Health Hazards of VDTs?* John Wiley & Sons, Chichester (1984).
213 E. van der Zee and A. W. van der Meulen: The influence of field repetition frequency on the visibility of flicker on displays. *IPO Annual Progress Report*, **17**, 76–83 (1982).
214 P. Zipp, E. Haider, N. Halpern and W. Rohmert: Keyboard design through physiological strain measurements. *Applied Ergonomics*, **14**, 117–122 (1983).

# *Index*

Abduction (of arm or of hand)   113, 115–117, 146
Abortions   202, 204
Accommodation   14, 19–21, 59–61, 73
Accounting machine operator   109–113
Adaptation (to darkness, to light)   23
Adjustable backrest   130–133
Adjustable seat height   133
Adjustable VDT workstation   135, 141, 142, 147–149
Adrenalin   167, 172, 173, 175, 176, 186
Adverse pregnancy outcomes   202, 203, 205
"Almost daily pains"   110, 111
Airborne dust particles   199
Angle of flicker perception   70
Anecdotal reports   202
Annoyance   161
Anthropometric data   101, 103, 143
Anthropometry   101
Antireflective devices   75, 82–86
Arm   107, 108, 110, 112, 113, 138
Arrangement of work surfaces   154, 155
Auditory nerve   160
Autonomous working group   187

Back   103, 108, 116, 117, 121
Backache (back pain)   103, 108, 116, 121
Background luminance   91
Background noise   160, 161, 164
Back load   116
Back muscles   103
Backrest   103, 105, 116, 129–132
Backrest angle   129, 132, 134
Backrest inclination (declination)   126, 133, 134
Birth defects   202, 204, 205
Blood pressure   167, 175, 176
Blood supply   97
Blurred characters   64, 73–75
Blurred vision   55, 58, 60
Body length (or size)   101, 103

Boredom   170, 171, 181, 186, 188
Brightness (contrast)   32, 41, 42
Burning eyes   55, 58, 60
Buttocks   105, 130

CAD workstations   9
Capital letters   87
Cataract   200
Cataractogenic effects   201
Catecholamines   167, 170, 172, 175, 184, 185
Cathode ray tube (CRT)   10
Cervical spine   116, 126, 127
Cervical syndrome   127
Character colour   14, 30, 31
Character contrast   30, 76–79, 86, 87
Character luminance   13, 77, 78
Character recognition   29
  size   86, 87, 89
Chromatic aberration   14
Ciliary muscle   17, 20, 21
Coloured filters   83, 85, 86
Colours   42, 43, 46
Complexity (of a job)   169, 185
Computer-aided design (CAD)   7–9, 56, 57, 109, 110, 147, 148
Computer breakdowns   174, 177, 184
Cones   18
Constrained postures   96, 99, 135, 150
Contrast ratio   91, 94
Contrast sensitivity   26, 27, 61
Control groups   56
Convergence (vergence movements)   25, 60, 61
Conversational jobs   7, 40, 43, 47, 109, 110, 112, 151, 155, 176
Cooling fans (noise)   166
Cornea   17, 18
Critical fusion frequency (CFF)   34, 35, 59, 60, 69, 70, 92
Curtains   53

Daisy wheel printer (noise)   165

223

# Index

Data acquisition (or retrieval)  6, 176–178, 180, 184, 188, 189
Data entry  6, 40, 56, 57, 109–112, 154, 171, 173, 175, 176, 178, 180, 184, 188, 189
Day-dreams  172
Dazzle  24
Desk height  103, 105
Desk surface  155
 *see also* Inclined desk top
Decibel  157, 158, 159
Deltoid muscle  104, 105
Descenders  89
Disc (load)  117, 118
Disc injuries  121
Disturbed linearity  79, 80
Dot matrix  11, 88–91
Drift  79, 80
Duration of gaze  8
Dynamic effort  96, 97, 99, 107

Effects of noise  160
Elbow angle  114, 115, 146, 149
Elbow height  101, 104, 106
Electrical activity of muscles (EMG)  104, 118, 123, 125, 136, 139, 140, 151
Electrical typewriter  107
Electron beam  10, 12, 14, 78
Electromagnetic radiation  193–195
Electrostatic field  196–200
Electrostatic potential  197, 198, 200
Equivalent noise level, $L_{eq}$  159, 162, 163, 164
Erect (upright) sitting posture  122, 123
Etching (the screen glass)  83–86
Eye blinking rate  27, 41
Eye complaints  37, 55, 57, 60, 62
Eye drops (use of)  64
Eye level  101
Eye movements  17, 25
Eye rotation  128
Eye saccades  28, 29
Eye strain  35, 56, 57, 62

Factor analysis (visual discomfort)  60
Field studies (surveys)  55, 56, 59, 108, 131
Filament lamps  34
Fixation of eyes (gaze fixation)  29, 30
Flat keyboard  151
Flicker (flickering light)  12, 34, 35, 64, 70–72, 92, 94
Fluorescent tubes (fluorescent lighting)  34, 35, 70, 71, 93
 phase-shifted  36
Focus  18, 19
Foetal abnormalities  203

Font  88–90, 92
Footcandle  32
Footrest  103, 105, 133
Forearm (support)  115–117, 137–139, 142, 144, 151–153
Fovea (centralis)  17, 18
Fragmented work  171, 186, 189
Frankfurt plane  129

Genetic effects  202
Glare (relative, absolute, adaptive)  24, 40, 47, 63, 93
Glare (sources)  37, 47, 50, 52, 93
Glare shields  48, 53

Hand (IO)  110, 113, 117, 137
Hazeltine font  90
Head  112, 116, 129
Head inclination  113, 114, 116, 127
Health complaints  178, 179
Health standards  195, 196
Hearing losses  160
Heart rate  167, 175, 176, 185
Height to width ratio of letters  79, 87
High-pitched noise  161
Home row  151
Horizontal leg room  106, 150
Huddleston font  88, 89

IBM font  90
Illumination (level)  32, 36–40, 91–94
Incipient cataract  200, 201
Inclined desk top (sloping desk)  118, 119
Indirect lighting  53
Indoor noise  162
Inflammation  100
Information age  2
Infrared radiation  193, 195, 201
Inkjet printer (noise)  165
Inner ear  160
Interactive jobs  7
Interline distance  30
Intervertebral disc (load)  117, 118, 120–122, 124, 126
Intervertebral disc pressure  121, 122, 124, 125, 152
Intervertebral disc troubles  100
Inward rotation of forearm  150
Ionizing radiation  193–195, 201
Iris  22
Itching eyes  55, 58, 60

Jitter  79, 80
Job broadening  187, 188
Job content  179, 189
Job control  168, 188
Job design  185, 186, 189
Job distress or dissatisfaction  169, 176, 177, 181, 184, 186

## Index

Job enrichment  187, 189
Job redesign  188
Job satisfaction  59, 170, 172, 174, 176, 177, 182–185, 189, 190
Job security  169, 174

Keyboard  150, 153
Keyboard desk  142
Keyboard level above desk  142, 153
Keyboard level above floor  115, 137, 141–143, 149
Key displacement  153
Key strokes, number  111, 113
Keys  151
  resistance  153
Knee  103
Kneeling posture  123
Kyphosis  117, 118, 122, 123

Laboratory experiments on preferred settings  135–137
Laser printer (noise)  165
Lateral abduction (ulnar) of hands  150–153
Leg room, vertical  106, 116, 150
Legibility  64, 67, 86–88, 90, 92
Lens  17, 18
Lens opacities  200, 201
Light fixtures  52, 53
Lighting (direct and indirect)  33
  temporal uniformity  46
Lights (appropriate, unsuitable)  47, 48, 49
Line of sight  127–129, 138
Lincoln-Mitre font  88, 89
Localized fatigue  98, 99
Lordosis  122, 123, 127
Loudness  158
Louvers  53
Lumbar pad (support)  126, 131, 133, 134
Lumbar spine (area or region)  116, 122, 123, 132
Luminance (surface luminance)  32, 42, 43, 71
Luminance contrast (contrast ratios)  40–45, 63, 67, 76, 77
Luminance oscillation  12, 36, 64, 65, 67, 69
Luminaries  52
Luminous flux angle  53
Lux  32

Malformations  202, 204
Man–machine system  4
Matrix printer (noise)  165
Mechanical typewriter  107
Medical findings  111, 112
Mental fatigue  175
Micromesh filters  82–86
Microwaves  193, 196, 201

Miscarriages  203, 205
Mixed jobs  188
Modulative transfer function (MTF)  67, 68, 74
Monotony  170, 177, 184, 188
Mood  168, 170, 175, 178, 179, 181, 184
Moulded chair  131
Movable keyboard  117, 153
Multipurpose chair  130
Muscular fatigue  98
Musculoskeletal troubles (discomfort)  100, 103, 105, 107–109, 112, 138, 146, 181

Near point  20, 21
Near vision  20
Neck (pain or discomfort)  103, 107, 108, 110, 112, 113, 116, 127, 135, 138
Neck–head posture  127, 129, 146
Neck pains (troubles)  114–116, 127, 135
Neck stiffness  114, 115, 135
Negative presentation of characters  14, 92
  see also Positive presentation of characters
Neutral density filters  82
Noise
  background  159
  cumulative frequency of
    levels  34, 35, 59, 60, 69, 70, 92
  emission (of office machines)  165
  level  157, 161, 162, 165
  load  159
  peak levels  162–165
Non-ionizing radiation  193, 194
Noradrenalin  167, 175

Occupational dermatitis  199
Occupational stress  167, 168, 180, 184
Office chair  130–133
Office desk (without typewriter)  106
Optic nerve  16–18
Oscillating luminance (light)  12, 13, 35, 63, 69, 70
Oscillation degree  70–72
Oscillogram  13
Outdoor noise  162

Painful pressure points  111, 112
Palpation findings  111, 112
Partial adaptation  24
Pelvis  122, 123
Perinatal mortality  202, 205
Persistence of phosphor  12, 69
Person–environment fit (PEF)  169, 170

Phoria (heterophorias)   60, 61
Phosphor (compounds)   13
  decay time   12, 69, 72, 92
  layer   10
Photometric characteristics   65, 66, 67
Photomultiplier   66, 67
Physical discomfort   107–111, 113–115, 128, 129, 134, 138, 140, 146, 147, 150
Pitch   157, 158
Polarization filters   84, 85
Positive presentation of characters   13, 14, 92, 94
  see also Negative presentation of characters
Posture changes   128
Preferred colours of characters   14
Pregnancy   202–204
Prematurity   202
Presbyopia   21
Pressure load on support   138–140, 152
Prismatic pattern shield   53
Psychosomatic disturbances   167, 173, 177, 178, 184
Punch card machines (noise)   163
Pupil, pupil aperture, pupil size   16, 17, 22, 24

Quarter-wave coating   84–86

Radio-frequency radiation   193, 196
Raster scan CRT   10
Readability   63, 86, 87, 90, 92
Reading (performance)   28–30, 62, 63
Recommendations   206–209
Reflectance   33, 42, 43, 45
Reflections   48, 50–52, 64, 82–86, 92–94
Refresh rate   12, 69, 70, 92–94
Repetitive task (work or job)   170–173, 183, 184, 186, 188, 189, 190
Reproductive hazards   202, 205
Response time (of the computer)   174, 175
Resting accommodation   20, 61
Rest luminance   78, 79
Rest pauses   190–192
Retina   17, 23, 24
Reversed video   13, 69–71, 91–95
Rods   18
Roughening (the screen glass)   83, 85

Safety standards   195, 196, 201
Scan lines   10, 11
Screen angles   137, 141, 143, 148, 149

Screen background (luminance)   77
Screen distance   137, 141, 149
Screen height   128, 137, 141, 143, 148, 149
Seat angle   124, 125
Seat height   137, 143, 148
Seat profile   130
Seat–desk distance   105
Sensory impulses   16
Sharpness of characters   65, 67, 73–75, 77, 92, 94
Shoulder (discomfort)   103–105, 107, 108, 110, 112, 113, 135, 138, 151
Shoulder pains   114, 115, 135
Shoulder stiffness (cramps)   114, 115, 127, 135
Sideward twisting   150–153
Sitting behaviour   130, 131
Sitting machine   130
Sitting posture   102, 119, 122–125, 130–134, 145
Skin disorders (troubles)   199, 200
Skin rashes   198, 200
Small letters   89
Social contacts   187, 189
Social support   169
Sound conditioning   164
Sound frequency   157
Sound perception   157
Sound pressure   157, 160, 161
Source document   154, 155
Source of noise   161
Spaces between letters   79, 87
Spaces between lines   79, 87–89
Spatial balance of brightness   40
Speech communication   160
Spine   121
Split keyboards (halved keyboards)   151, 152
Spontaneous abortion   202–205
Stability of characters   65, 79, 80, 81
Static effort   96–99
Static load   100
Static muscular work   96, 98, 99, 105
Static posture   99
Status symbol   119, 120
Still-birth   203
Stress   167, 168, 170, 172, 174, 178, 183–186
Stressors (psychosocial)   168, 169, 178, 181, 183, 184, 186, 188
Stroke width   87, 88, 92–94
Subharmonic oscillations   35
Syntop chair model   134

Tablet   8, 9
Task lighting   53
Taylorism   171, 186
Telephone operators   57, 58, 108

## Index

Teratogenic risk (hazard)  205
Threshold of audibility  158
Tiltable chair  130, 131, 133
Trapezius muscle  104, 105, 118, 123, 136, 138, 140
Tremor of the eye  25
Trunk posture (inclination)  144–146, 148, 149
  upright  103, 107, 117, 118, 123, 124, 148, 149
Typewriter  104, 105, 150, 151, 163, 166
Typing desk  106
Typist (fulltime typist)  56, 57, 109, 110–112

Ultraviolet radiation  193, 195, 201
Unnatural postures (position)  121, 150
Upper and lower case letters  89
Upper limbs  118, 125, 138, 139
User-friendly controls  133, 150

Van Nes font  90, 91
VDT jobs  6
VDT workstations,
  positioning  52
  preferred settings  135, 143, 147–149
Viewing angle  128–130
Visible flicker  12, 14, 35, 63, 64, 94
Visual acuity  22, 26, 59, 60, 61, 200–202
Visual capacities  25, 26, 60
Visual comfort  67

Visual deterioration  200
Visual discomfort (complaints)  35, 55, 57–60, 63, 65, 181
Visual distance (viewing distance)  105, 113, 124, 143
Visual down angle  143
Visual fatigue  21, 55, 60, 67, 181
Visual field  18, 19, 48
Visual perception, speed of  16, 17, 28
Visual performance  47
Visual reading field  30
Visual scanning  7, 8
Visual tests  60, 61, 62
Voice  160, 161

Weighted sound level (dB(A))  158, 159
Window  52, 53
Word processing  7, 173, 178, 180
Work efficiency  185
Working height  102, 105, 106
Working hours (at VDTs)  190
Work load  174, 175, 185
Work motivation  185
Working posture  113
Work-sampling method  8, 9, 103
Work variety  187
Wrist (support)  115, 116, 137–139, 142, 144, 151–153

X-ray radiation (emissions)  193–195

Yellow-green characters  14